Caring for Small Woods

For Norma, Wayne, Adrian and Justine

Caring for Small Woods

A practical manual for woodland owners,
woodland managers, woodland craftsmen,
foresters, land agents, project officers,
conservationists, teachers and students

Ken Broad

Earthscan Publications Ltd, London

First published in the UK in 1998 by
Earthscan Publications Ltd

A catalogue record for this book is available from the British Library

ISBN: 1 85383 454 8

Typesetting by PCS Mapping & DTP, Newcastle upon Tyne
Printed and bound by Biddles Ltd, Guildford and Kings Lynn
Cover design by Yvonne Booth

For a full list of publications please contact:

Earthscan Publications Ltd
120 Pentonville Road
London, N1 9JN, UK
Tel: +44 (0)171 278 0433
Fax: +44 (0)171 278 1142
Email: earthinfo@earthscan.co.uk
WWW: http://www.earthscan.co.uk

Earthscan is an editorially independent subsidiary of Kogan Page
Limited and publishes in association with WWF-UK and the
International Institute for Environment and Development.

Contents

List of Figures, Tables and Boxes

FIGURES

TABLES

BOXES

List of Photographs

Acronyms and Abbreviations

ADAS	Agricultural Development and Advisory Service
AMAAA	Ancient Monuments and Archaeological Areas Act 1979
AONB	Area of Outstanding Natural Beauty
APF	Association of Professional Foresters
ASSI	Area of Special Scientific Interest (Northern Ireland)
BBONT	The Bucks, Berks and Oxon Naturalist Trust
BOAT	byway open to all traffic
BTCV	British Trust for Conservation Volunteers
BTMA	British Timber Merchants Association
CLA	Country Landowners Association
CPRE	Council for the Protection of Rural England
dbh	diameter at breast height
DIY	do-it-yourself
ECCTIS	Educational Counselling and Credit Transfer Information Services
EN	English Nature
ESA	Environmentally Sensitive Areas
ESUS	East Sussex Small Woodland Project
ETSU	Environment Technology Support Unit
FA	The Forest Authority
FASTCo	Forestry and Arboricultural Safety and Training Council
FC	Forestry Commission (comprises Forest Enterprise and the Forestry Authority)
FCA	Forestry Contracting Association
FE	Forest Enterprise
FICGB	Forest Industry Committee of Great Britain
FRCA	Farming and Rural Conservation Agency
FSC	Forestry Stewardship Council
FWAG	Farming and Wildlife Advisory Group
FWPS	Farm Woodland Premium Scheme
FY	forest year

ha	hectare (1 ha = 2.471 acres)
HGTMA	Home Grown Timber Merchants' Association
hp	horsepower
HSE	Health and Safety Executive
ICF	Institute of Chartered Foresters
LNR	Local Nature Reserve
MAFF	Ministry of Agriculture, Fisheries and Food
NCC	Nature Conservancy Council
NFU	National Farmers Union
NNR	National Nature Reserve
NR	natural regeneration
NRA	National Rivers Authority
NSWA	National Small Woods Association
ob	over bark
OS	Ordnance Survey
pH	A measure of acidity and alkalinity – used in forestry primarily to test soil reaction; it operates on a scale of 0–14. A pH of 7 is neutral; less than 7 is acid; and more than 7 is alkaline.
Pyr	planting year
PTO	power take-off
RFS (EWNI)	Royal Forestry Society (of England, Wales and Northern Ireland)
RoW	right of way
RSPB	Royal Society for the Protection of Birds
RUPP	road used as a public path
SRC	short rotation coppice
SSSI	Site of Special Scientific Interest
TAP	Training Access Points
tdob	top diameter over bark
tdub	top diameter under bark
TGA	Timber Growers Association
TPO	Tree Preservation Order
ub	under bark
VAT	value-added tax
WGS	Woodland Grant Scheme. A Forestry Authority scheme of grant aid for the establishment and management of trees and woodlands.
WT	Woodland Trust

Foreword

I have known Ken Broad for a number of years and in that time have watched with admiration the growth of the Oxfordshire Woodland Project which he manages.

This practical book, aimed at the non specialist, is a distillation of Ken's lifetime's experience as a forester, both with that project and with the Forestry Commission.

In Ken's own words, he has not duplicated the wealth of knowledge available in many books on, for example, woodland nature conservation. Instead he has provided a guide to *Caring for Small Woods* full of useful information including the many figures, tables and boxes that enable information to be quickly accessed.

Writing a book on such a vast subject is not an easy task, but in deciding to write one of a practical, rather than a theoretical nature, he has been clear about what he set out to do and this is the book's strength. The information is presented in a way that leaves the reader to decide on issues such as control of pests and diseases and sporting, without getting bogged down in all the arguments.

This book is based on Ken's everyday work having to meet owners of woodland and help them to balance their objectives with the realities of the site, the timber market and so forth. In the National Small Woods Association (NSWA) we receive a large number of enquiries from people wanting advice on how to buy and manage small woods. This book is very timely, not least because interest in small woodlands has exploded in the last five years. A lot of basic information on small woods is spread through many unrelated publications. *Caring for Small Woods* helps in that it can also be used as a reference guide that can lead the reader into more research on an individual topic, such as management planning, which can be a whole book in itself.

A famous Chinese philosopher, Lao Tzu, one said that 'a journey of one thousand miles begins with a single step'. In NWSA we help people interested in woodlands to make that first step,

3, Perkins Beach Dingle
Stiperstones,
Shropshire SY5 0PF
Tel: 01743 792644
Fax: 01743 792655

Chairman: Tony Philips OBE

Executive Director: Russell Rowley

Company limited by guarantee no. 3390162

Our mission is to "improve or maintain the productive, aesthetic and environmental value of the nation's small woodlands by sustainable management through education, training, development markets and improving information exchange".

If you own a woodland, are thinking of buying one, or are just interested, then:

- *Join the leading UK organisation specialising in Small Woodlands*
- *Receive HeartWOOD, our quarterly magazine full of interesting articles*
- *Meet others at our annual conference, events and training days all over the UK*
- *Have access to a free telephone help line on all aspects of woodlands*
- *Get low cost woodland insurance*
- *Receive discounts of up to 20% on woodcraft courses*
- *Help us lobby through your membership to prevent the ongoing neglect of small woods. For over a decade we have been the voice of small woods and have raised their profile enormously, commenting constructively on documents from government*
- *Find out about a scheme to own woodland with others*

The National Small Woodland Association (NSWA) was formed in 1988, and is the leading independent organisation in the UK working solely with small woodlands. Over half the total area of broadleaved woodland in Britain is made up of small woodlands, and well over half of this total has received little or no management over the last thirty years. All woodlands need managing to thrive, to produce timber, and to be good for wildlife. In NSWA we specialise in offering a personal, friendly service that cuts through the jargon and tells you what you need to know.

TO JOIN, WRITE OR PHONE ROGER PITTAWAY, MEMBERSHIP MANAGER, FOR MORE DETAILS.

and if this book helps you to have as enjoyable a journey with small woods as I have had, whets your appetite, and helps you take that first step, I am sure that Ken will feel that all his work has been worthwhile.

Russell Rowley Msc (Env For) MIEEM
Executive Director
National Small Woods Association

Preface

Caring for Small Woods is meant to provide owners with the practical help and encouragement they need to improve their small woodlands. A small woodland is generally accepted as one having an area greater than a quarter of a hectare but less than 10 hectares. Nonetheless, the advice given in this book is more to do with the way people relate to woods and their management than it has with any size criteria. The emphasis throughout is on the management of existing woods rather than the creation of new ones.

A basic philosophy underscores the book. For any programme of work to succeed it needs to be recognized that the most important player is the woodland owner. I am of the firm conviction that owners must take responsibility for managing their own woods; without their enthusiasm, dedication and personal involvement, there is little chance of developing that long-term commitment to woodland management that I consider to be absolutely essential.

That is not to imply that the book is for woodland owners alone. *Caring for Small Woods* is meant to promote, to a much wider audience, a greater understanding of the need to manage small woods. It is for woodland managers, consultants, contractors, land agents, foresters, craftsmen, farmers, farm managers, teachers, students and conservationists. It should serve as a useful work of reference for fieldworkers, local authority officers and woodland project staff. It is for those who work part-time in woodlands, those who may work full-time but whose responsibilities are limited, and those who simply enjoy being in woods and wish to understand them a little better.

This is not, however, another book about ancient woodlands, though it focuses on one of the surest ways of conserving them – namely caring for them in responsible ways. It is not a guide to tree recognition – there are already enough good books on that subject to fill a fair-sized bookcase. And it is not a treatise on economic forestry – there is nothing here on production forecasting, investment appraisal, budgeting or projected returns; the

Forestry Authority has produced scores of publications for study on such matters. Indeed, the book contains little to challenge the intellect. Rather, my aim has been to sift and interpret the important facts regarding the day-to-day management of small woods, and then to present them in a form that can be easily digested and applied by the non-specialist.

Woodland management terms, jargon and abbreviations, freely used by professionals and specialists throughout the industry, are not always fully understood by the novice. In most circumstances, a lack of familiarity with technical and trade terms will be of no great disadvantage to those who care for small woods on a practical level; however, where grant schemes, felling licences and other legal agreements are concerned, the particular meaning of various words and phrases can be more significant. I have made it an aim in writing this book to avoid the use of jargon as much as possible, but if the reader does experience any difficulty on that score then reference to the comprehensive glossary at the end of this book should clear up any confusion. There is also a comprehensive list of addresses which should be useful if further information on any particular topic is required (see Appendix 4).

Since the book is written primarily with the novice in mind, the conventional but somewhat distracting practice of inserting authority references within the text has been avoided. Reference to what I have read, and found useful, is by way of superscript numbers with endnotes, together with a list of further reading in Appendix 5.

There is a growing awareness that woodland management is a good thing. People do care about what happens to small woodlands. But caring about them is not enough; caring *for* them is what matters. This is what *Caring for Small Woods* aims to address.

Ken Broad
Haddenham, Buckinghamshire
June 1997

Acknowledgements

Having spent the best part of a lifetime caring for state-owned woodlands and providing management advice for private woodland owners, it would be unusual, to say the least, if I had not learnt something about the care of woodlands through my own endeavours. That said, I readily acknowledge that there is little in this book that has not been said or written before and I concede to having consulted a large number of authorities in its preparation.

The book evolved from a series of information leaflets that I prepared as manager of the Oxfordshire Woodland Project between 1991 and 1997. I am especially grateful to Eric Dougliss, Oxfordshire County Council forester, for editing most of those leaflets and for allowing me to use extracts from some of his unpublished papers. I also profited greatly from his comments on an early draft of this book.

The information leaflets referred to would not have been published without the generous support of many sponsors, including:

- Castrol International;
- English Nature;
- Fountain Forestry;
- Wessex Woodland Management Ltd;
- Correx Plastics;
- Lonsdale Forestry;
- Woodland Improvement and Conservation Ltd;
- Acorn Planting Products;
- Cherwell District Council;
- National Rivers Authority (NRA);
- Marlwood Ltd;
- The Oakley Wood Group;
- Southern Forestry;
- Chantler Timber;
- The Forestry Authority;

- Wessex Timber;
- BBONT (the Bucks, Berks and Oxon Naturalist Trust).

I am indebted to the Forestry Commission for permission to use information from various publications. Thanks are due to Richard Wise, graphic designer, for allowing me to reproduce some of the illustrations that he prepared for the said information leaflets. I am also grateful to Iain Corbyn of BBONT for comments on wildlife conservation; Jeremy Biggs of Pond Action, Oxford (Brookes) University, for advice on woodland ponds; James Andrews of Marlwood Ltd for advice on mobile sawmills; Colin S Cooper, Read Cooper Solicitors of Thame, Oxfordshire, for advice on conveyancing; Patrick Stephens, Forestry Authority operations manager, on whose early draft the glossary is based; Daphne Fisher, for research on pollarding literature; Jonathan Howe, for advice on coppicing; the National Small Woods Association, for permission to use extracts from the *Woodlands Initiative Register*; Geoffrey Sinclair, Woodland Officer, Suffolk Country Council, for comments on the list of woodland initiatives.

For general comments I must also thank Roger Wills, independent woodland management consultant; Rick Pakenham, Chiltern Forestry; and Malcolm Otter; Ridgeway national trail coordinator.

Introduction

THE IMPORTANCE OF SMALL WOODS

No precise information is available as to the size and nature of Britain's small woodland resource, but in 1992 there were thought to be around 200,000 small woodlands under 10 hectares (ha) with a combined area of about 500,000 ha in the management of some 100,000 owners.[1] The area occupied represents between one quarter and one third of the country's total woodland. In England, about 90 per cent of woodlands are small woods. And small woods continue to be planted in significant numbers; in 1995–96, over 90 per cent of new planting schemes which came under Forestry Authority Woodland Grant Schemes involved areas under 10 hectares. The average area was 2.1 hectares.[2]

The benefits provided by well-managed small woodlands are many but are seldom fully appreciated. The most important are summarized here.

Timber and Wood Production

There is a huge volume of standing, low-grade material in the undermanaged small woodlands of Britain. Since the majority of these woods are found on better soils than those occupied by traditional forestry, there is ample potential for improvement.

Import Substitution

Approximately one million cubic metres of hardwood sawn-timber are used in Britain each year. Approximately 70 per cent of this is imported and about half of that comes from the tropics.[3] By producing more British sawn hardwood we have the potential to reduce pressure on woodlands in other parts of the world.

Landscape Enhancement

Small woods are important landscape features. Well managed, they enhance the beauty and character of the countryside and contribute to maintaining the diversity of rural scenery.

Nature Conservation

Small woods can be important for wildlife. Britain's woodlands are considered to be among the richest of ecological sites.

Recreation

Small woods provide opportunities for a range of recreational activities. Although access to the countryside in Britain is regulated, there is a vast network of footpaths and bridleways, and many pass through small woodlands.

Therapeutic Properties

Woods provide ideal places in which to relax for those engaged in stressful employment, or in which to recuperate for those recovering from illness.

Amusement

Many owners manage their woods simply as a hobby.

Educational Facilities

Small woods can provide educational opportunities by establishing nature trails, arboreta and school plots.

Shelter

Woodlands play an indispensable role in the provision of shelter: shelter for people, homes, crops and livestock.

Screening

Groups of trees can be used to screen unsightly buildings, to help conceal car and caravan parks or to screen a house from a main road.

Carbon Fixing

Trees absorb carbon dioxide from the atmosphere, helping to reduce global warming. Throughout their growing lives, trees store the gas as carbon. As trees rot and die the carbon is returned to the atmosphere as carbon dioxide. But if trees are felled and utilized, the carbon is retained in the timber until destruction.

Job Creation

Small woodlands can provide work and create opportunities for economic diversification in the countryside.

Property Values

Woodlands can increase the value of rural properties.[4] The presence of well-established woods makes the land more desirable and extends the range of potential buyers.

Tourism

The image of a whole region can be enriched by well designed and well-landscaped woodlands, and this can enhance tourism.

Water Quality

Woods alongside rivers, streams and ponds exert a fundamental influence over the health and productivity of the freshwater ecosystem.

Heritage Value

Small woods are part of our history and our heritage. What happens to them is vitally important.

SMALL WOODS IN CRISIS

The majority of small woods in Britain are broadleaved woods, which means they are relatively slow growing. Their small size means they lack the economies of scale. Furthermore, access is often difficult, owners generally lack traditional skills, markets for small woodland products no longer exist in any volume and many woods now reflect centuries of negative selection under which the best trees were 'creamed off', leaving a stock of inferior quality trees. These, and other factors, have led to the widespread neglect of small woodlands and their present condition is giving cause for concern.[5] By some estimates 80 per cent are unmanaged or under-managed. Worse, there is a widespread belief that woods are actually better left unmanaged. For small woods nothing could be further from the truth. They are fragile and vulnerable. Most require some management if they are to survive and prosper.

Unmanaged woods, far from being havens for wildlife, are more likely to be just the opposite. The Forestry Authority, English Nature, the equivalent nature conservation authorities in Scotland and Wales, and many, if not all, county naturalist trusts agree that managed woodlands are more likely to survive than unmanaged ones. Indeed, our woodlands are much more the products of past management than they are natural features. Most ancient woodlands have survived because they were managed, and some may be richer in wildlife than the natural woods from which they were descended. But management is expensive, so income from woods is crucial. In most cases, small woods can be made to pay their way without impairing landscape, wildlife or recreational interests, all of which are better served by healthy trees than by ailing stock. This is obviously in both the owners' and the national interest.

Woodland Neglect

Neglect of woodland, on the scale we see today, is a relatively recent phenomenon. Until about 100 years ago, almost all woods were managed for one reason or another. They constituted a vital part of the rural economy, providing timber and wood for:

- ship building;
- house building;
- furniture;
- mining;
- wagon and cart construction;
- farm implements;
- cart wheels;
- fence material;
- sheep hurdles;
- hop poles;
- thatching spars;
- tent pegs;
- barrel hoops and staves;
- charcoal;
- turnery;
- bark for tannin;
- pea and bean sticks;
- clog bottoms;
- clothes pegs;
- and enormous quantities of fuel wood.

To produce such a variety of items, thousands of craftsmen were employed. Many worked outdoors in all weathers, either as individuals or in small units, some travelling daily from home, others wandering from place to place where they would set up temporary shelters of branches and turves. These hardy people included:

Thousands of small woods are in need of management

- copsemen;
- timber fellers;
- underwood cutters;
- horsemen;
- sawyers;
- bark peelers;
- chair bodgers;
- basket makers;
- hurdle makers;
- paling makers;
- clog makers;
- broom squires;
- rake makers;
- wheelwrights;
- hoop shavers;
- faggot makers;
- pole lathe turners;
- charcoal colliers.

Gradually, most of the traditional markets disappeared. Many old skills and customs were lost to industrialization and the drift to the towns. And the woods fell silent.

Previously, too, a country-wide network of small sawmills existed. They catered for the needs of local growers and wood users. But timber conversion has become highly automated and centralized, and sawmills are now far less accessible to the woodland owner. The cost of transporting timber half way across the country has the effect of discouraging many owners from managing their woods. Worse was to come. Apart from the adverse consequences of market decline and social change, in more recent times thousands of woods have been devastated by:

- the wholesale fellings during two World Wars;
- the deaths of over 20 million elms from Dutch elm disease;
- the catastrophic storms of October 1987 and January 1990;
- a succession of debilitating summer droughts – 1975 and 1976 were particularly dry years.

Fortunately, most of the woods felled during the First and Second World Wars have regrown. However, without management many have become dark and gloomy and are overstocked with tall, weak and spindly trees of mostly equal age. Other woods have become more open-grown, with shorter, coarse-branched trees and shrubs – so-called scrub woodland. Lack of management has resulted in broken fences, unlaid hedges, blocked culverts and drains, unstable trees that are easily damaged by wind, and overgrown paths and rides.

Woodlands can also deteriorate where they are used intensively for sporting. Gamekeepers generally regard any woodland work as potentially disruptive, and they can be surprisingly effective in persuading woodland owners to defer any operation that might disturb the pheasants.

WHY MANAGE WOODLANDS?

Britain imports 43 million cubic metres of timber – 90 per cent of its timber needs – at a cost to the nation of about £7 billion annually. This equates to about £850,000 every hour. Appropriately managed, our small woods could go a long way towards reducing this import bill. But the concept of management does not win universal approval. Why do woodlands need managing? Left to nature, some people argue, felled and damaged woods will grow again. When mature, trees will regenerate and multiply without the help of foresters. Furthermore, as time goes by, these woodlands should become more diverse as their structure alters and new species colonize. The net result should produce woodlands suitable for a wide range of wildlife and are attractive in the landscape.

Such is the theory. In practice, in many parts of the country, small woods do not regenerate naturally. And that means their long-term future is under threat. The situation seldom attracts attention because woodland degeneration happens so slowly. And far from becoming more diverse, many of these small woods have become featureless places. They often consist of a narrow range of tree species and lack anything that might be described as an understorey, making them cold and draughty.

The reason for the decline is seldom obvious. Most tree species regularly produce the seed needed for successful regeneration. Indeed, in mast years seed can be shed in such profusion that, in due course, the woodland floor will become carpeted with tiny seedlings. Yet hardly any of these little trees survive long enough to produce replacements for the mature trees that will eventually grow old and die. (There are exceptions – groups of young ash are a common sight and sycamore can colonize a site so competitively that it may shade out most other trees.) But many species do not survive much beyond the first year. The reasons for this failure are not fully understood but the following factors are thought to be important.

Insects

Predation by insects can begin even before seeds are ripe or are shed.

Infertility

Not all seed is fertile, and perhaps modern pesticides and fertilizers are having an adverse effect – this could be important where farm woodlands are concerned.

Birds

Finches feed on the seeds of many tree species. Pigeons eat enormous quantities of fallen acorns. They can take 50 to 60 in a single full crop.

Animals

The browsing and grazing activities of animals, both domestic and wild, are a great problem. Tree seed is an important autumn food for some rodents. Woodmice will eat freshly germinated seedlings. Voles nibble larger seedlings. Grey squirrels eat acorns and hazel nuts and will nip the tops off seedlings. Where numbers are high, rabbits, hares and deer can cause considerable damage if they exceed the carrying capacity of the wood.

Extremes of Climate

Prolonged low temperatures, early and late frosts, frost lift, heat and moisture stress can wipe out a whole season's regeneration.

Disease

Several seed-borne or soil-borne fungi can attack young roots and stems. They usually occur in the first three to four weeks after germination. The condition is called damping off.

Lack of Light

Young trees will perish unless they receive sufficient light, since they are adapted to colonize open spaces or areas of dappled shade. Most seedlings will fail in the dense shade cast by sycamore and beech trees, while species such as elderberry, Japanese knotweed, snowberry and bracken can colonize land to the exclusion of almost all other growth. These are just some of the factors that may influence the survival of young trees. There are probably others.

THE RESTORATION OF SMALL WOODS

It is unwise to assume that, just because they are of modest size, the rehabilitation of small woods must be a simple business. It is not always so. Small woods have their own intrinsic problems, and raising unrealistic expectations amongst their owners can lead to disappointment and may even act as a disincentive to management. One of the main obstacles is the difficulty of focusing attention on the plight of small woods. Schemes involving new planting are always popular but they can divert attention away from the condition of existing woods. And although it is easier to create productive woodland on green field sites than it is to rehabilitate old woods, it makes little sense to do so while so many woods lie derelict.

Small woods can present other problems:

- Woodland improvement is rarely the owner's main interest.
- Owners often lack the necessary skills to do the work themselves.
- Owners may also lack the confidence to supervise contractors.
- In spite of the will and the ability to manage, owners frequently lack clear objectives.
- Costs in small woods are disproportionately high and returns disproportionately low.

- Marketing timber from small woods can be difficult.
- Where timber production is an important objective, quality must be high.
- Even where timber quality is high, at least one lorry load is normally required to produce a reasonable profit.
- Access is often difficult – small, potentially unprofitable woods are seldom well roaded.

ADVICE AND ASSISTANCE

The difficulties associated with small woodland management are becoming much better understood. Care in the management of the country's small wood resource now figures prominently in the objectives of a number of organizations. In some regions, the number of organizations involved in the provision of small-woodland management advice is now so large as to be counterproductive. Owners are unsure who best to approach. Help is on hand from government and non-government organizations, community and urban forests, trusts, charities, groups of many kinds and a host of national, regional and local woodland initiatives. Some of these bodies will give advice free of charge, others may levy a fee. Generally, the smaller the area of woodland and the less profitable the prospects of management, the more likely the probability of obtaining a free service. In the first instance it is worth contacting the local authority or the regional Forestry Authority office. They are likely to have an idea of what is available locally.

The key players and supporters who have been responsible for ensuring that the needs of the small woodland owner are recognized are detailed below.

The Countryside Commission

The Countryside Commission has played a leading role in identifying solutions to the problems of widespread deterioration and neglect in small woodlands.[6] In previous years the commission funded a number of woodland initiatives in various parts of the country. Most of these projects offered free advice and other services to woodland owners. They have been effective in:

- raising the profile of small woods;
- reinstating management in thousands of them;
- stimulating the development of new markets for woodland produce; and
- creating a more positive climate among the owners of small woods.

Many local initiatives provide opportunities to learn about woodland management

The more successful projects have often been those that have been honest in telling woodland owners what to expect from their efforts. Many have been successful in raising levels of awareness of the wider benefits of woodland management. One of the most useful sources of information regarding tree and woodland organizations is the Woodlands Initiative Register, produced for the National Small Woods Association. The register is meant to be updated at three-yearly intervals.

Initiatives understood to offer some free advice, and the areas they cover, will be found in Table I. The list is unlikely to be comprehensive and will be subject to change as new projects are set up and the less successful ones are wound down.

The British Trust for Conservation Volunteers (BTCV)

This organization provides advice to landowners and managers on suitable conservation projects and organizes groups of volunteers to carry out work. They operate through a network of local offices throughout England, Wales and Northern Ireland. Although the work will be undertaken by volunteers, a charge is made to cover costs. They can also supply materials. Once contact is made, a BTCV representative will meet to discuss requirements and assess the suitability of the project. Each project is negotiated individually and a written quotation is provided. Many of the projects they

Table 1.1 *Woodland Initiatives*

Project	Area of operation
Amman & Gwendraeth Valley Woodland Initiative	Mid Wales
Anglian Woodland Project	Cambridgeshire, Essex, Norfolk, Suffolk
Argyll Broadleaves	Argyll
Chiltern Woodland Project	Chiltern Hills
Cleveland Community Forest	Parts of Cleveland
Clyde Valley Woodland Initiative	Clyde Valley
Coed Cymru	Wales
Cumbria Broadleaves	Cumbria
East Derbyshire Woodland Project	East Derby
East Sussex Small Woodland Project (ESUS)	East Sussex
Forest of Avon	In and around Bristol
Forest of Belfast	Belfast
Forest of Cardiff	Cardiff
Forest of Mercia	South Staffordshire & Walsall
Great North Forest	South Tyne & Wear and North-East Durham
Great Western Community Forest	Swindon area
Greenwood Community Forest	Nottinghamshire
Gwent Small Woods	Gwent
Highlands Birchwoods	Scottish Highlands
Lancashire Woodlands Project	Lancashire
Marches Woodland Initiative	English–Welsh borders
Marston Vale Community Forest	Bedfordshire
Mersey Forest	Merseyside and North Cheshire
New Forest of Arden Project	Warwickshire
North-West Sutherland Native Woodlands	North-West Sutherland
North Pennine Woods Initiative	North Pennines
Oxfordshire Woodland Project	Oxfordshire
Pang Valley Charcoal	Berkshire
Red Rose Forest	Bury, Bolton, Wigan, Salford, Trafford and Manchester
Rockingham Forest Trust	Northamptonshire
Shropshire Forest	Telford Urban Fringe
Silvanus Trust	Cornwall, Devon, Somerset, Dorset
South Pennines Woodland Project	South Pennines
South Yorkshire Forest	South Yorkshire
Tayside Native Woodlands Initiative	Tayside
Thames Chase Community Forest	Essex and North London

Tir Cymen	North Wales
Valleys Forest Initiative	South Wales
Watling Chase Community Forest	South Hertfordshire
Wessex Coppice Group	South England
West Fife Woodlands Initiative	West Fife
Wildwood Project	South-East Hertfordshire
Woodland Wildlife Project	Hampshire

work on may be eligible for grant aid. They include:

- habitat management;
- restoration of hedges and dry stone walls;
- constructing footpaths, stiles, gates and bridges;
- woodland management;
- tree planting;
- creating and managing green space.

The Farming and Wildlife Advisory Group (FWAG)

This group was established to help farmers and landowners by providing expert advice and information on maintaining the essential balance between conservation and profitable agricultural operations. The FWAG specializes in providing practical whole farm plans.

Agricultural Development and Advisory Service (ADAS)

ADAS is an executive agency of the Ministry of Agriculture, Fisheries and Food (MAFF) and the Welsh Office. ADAS operates as two distinct sections:

- the commercial section, which has kept the name ADAS and is now an independent consultancy that charges for its services;
- the statutory section, now named the Farming and Rural Conservation Agency (FRCA), that carries out the technical aspects of MAFF's policy work and agri-environmental initiatives.

The National Small Woods Association (NSWA)

Established in 1989, the NSWA aims to promote the management, conservation and rehabilitation of small or undermanaged woods, through education, training, developing markets and improving information exchange.

In 1997 the Association became a limited company with the objectives of representing private and public owners of small woods

CHILTERN WOODLANDS PROJECT

Aims to promote and encourage the sustainable management of small woods across the Chiltern Hills. Much of this region is designated as an Area of Outstanding Natural Beauty and includes parts of Oxfordshire, Buckinghamshire, Hertfordshire and Bedfordshire.

The Chiltern Woodlands Project will:

- Give a FREE advisory visit to owners of unmanaged woods, with suggestions on woodland management and conservation, markets for timber, grants and felling regulations.
- Publish "News of the Woods" newsletters and annual reports.
- Act as regional co-ordinator for "Woodlots" for Bucks, Herts & Oxon.
- Organise woodland meetings and demonstration days.
- It can act as a consultant in preparing management plans and contracts, Woodland Grant Scheme applications etc. - a fee will be charged for this work.

For more information contact
John Morris, Chiltern Woodlands Project,
Pigotts, North Dean,
High Wycombe, Bucks HP14 4NF
Tel 01494 565749 or Fax 01494 565752

Registered charity no 1002512

The **Chilterns**
Area of Outstanding Natural Beauty

CHILTERN WOODLANDS PROJECT

The Chiltern Society formed its Small Woodlands Project in 1983, offering practical help to woodland owners across the Chiltern Hills. It was re-launched as the Chiltern Woodlands Project Ltd, an independent non profit making company in 1989 and became a registered charity in 1991. John Morris has been with the project since the start and has visited over 500 woodland owners in the Chilterns to offer advice and assistance.

This project aims to raise awareness and understanding of woodland management and conservation issues. It has organised a series of successful woodland meetings, events and talks, including a Charcoal Workshop. It offers much needed advice and encouragement to inexperienced woodland owners.

The Chilterns are known for its attractive wooded landscape, dominated by mature beech woods, which cover 20% of the land area. There are over 16,000 ha of woodland here, over half of these are ancient woods of considerable importance for nature conservation.

throughout Britain and supporting the many practitioners, from consultants to coppice workers to green woodworkers. A range of training and subject days has been planned and other membership benefits include a membership pack that explains the jargon, insurance offers, as well as advice and discounts on services. Their newsletter *Heartwood* keeps members in touch with what is going on, and provides the latest news and technical information for woodland owners, practitioners and those with a general interest in woodlands.

Although there is generally no shortage of free advice and assistance available to woodland owners, setting up contacts with owners is a major problem for those in a position to offer help. No central registry exists and not all owners live locally. There is a general assumption that most small woodland owners are farmers. In some places (East Anglia and Wales, for example) this is undoubtedly the case. But there are substantial differences in regional ownership patterns. For example, in the first three years of the Silvanus Project (in Devon and Cornwall) only half the clients were found to be full- or part-time farmers, while in a subsequent period this declined to 37 per cent.[7] In Oxfordshire, only about 40 per cent of the Woodland Project's clients turned out to be farmers; individual owners came from all walks of life, while corporate owners included churches, priories, district and parish councils, schools, Scouts, the Army Cadet Force, the county wildlife trust – and a naturist club.

Management Planning

Management planning provides a means of considering all the options available over a period of time. It ensures that the most important factors are given priority and it allows rational consideration to be given before any action is taken. To ensure continuity, plans need to be written down, and this remains true no matter how small the wood. This part of the book provides the guidance needed for drafting a small wood management plan.

Before preparing such a plan, however, an appreciation of a number of widely accepted guidelines, some fairly typical management options and a few commonly encountered constraints will be indispensable.

MANAGEMENT GUIDELINES

The aims of forestry, and of forest policy, have changed during the last 150 years. In the 19th century many woods were established and managed primarily for sport. After 1919, with the establishment of the Forestry Commission, large areas of land were given over to building a reserve of standing timber for use in times of war or national emergency, and private landowners were encouraged to do the same. Later, policies changed and the commission's main aim became the production of economic crops of timber.

Today, woodlands attract intense public attention. Managers and landowners are under pressure to provide benefits other than timber production. Access, attractive landscapes, recreation and wildlife conservation are in demand. However, conflicts of interests can arise (for example, where sporting and public access interests overlap). It follows, therefore, that the production of an effective management plan is a fundamental requirement of woodland planning.

Broadleaves Policy

Concern about the continuing loss of ancient woodlands contributed to the government's decision, in 1985, to introduce the Broadleaves Policy. This aims to maintain and increase broadleaved woodland by encouraging good management for a wide range of objectives, giving special attention to seminatural ancient woodlands. This means the 'coniferization' of ancient woodland is no longer a valid option.

The Government's approach to forestry is defined in *The UK Forestry Standard* published by the Forestry Authority.[1] This sets out the criteria and standards for the sustainable management of all forests and woodlands, state-owned and private, in the UK. The standard is linked to the developing international protocols for sustainable forestry. On the principle that each situation is unique and demands individual attention in order to develop appropriate plans and working practices, the publication encourages owners to adopt the methods appropriate to their objectives by means of a set of six Standard Notes:

- General Forestry Practice;
- Creating New Woodland;
- Creating New Native Woodland;
- Felling and Restocking Planted Woodland;
- Managing Semi-Natural Woodland;
- Planting and Managing Small Woods.

The Forestry Authority has published a comprehensive set of advisory guides on the management of seminatural ancient woods throughout Britain that take account of local and regional factors and that describe the management most appropriate for each type of woodland. While the guides are aimed at seminatural ancient woodland, much of the advice will also be appropriate for other seminatural woods that are of high conservation value, and for long-established planted woods that have developed some of the characteristics of seminatural ancient woodland. Eight broad woodland types are recognized:

- lowland acid beech and oakwoods;
- lowland beech-ashwoods;
- lowland mixed broadleaved woods;
- upland mixed ashwoods;
- upland oakwoods;
- upland birchwoods;
- native pinewoods;
- wet woodlands.

The guides are available from local Forestry Authority offices.

The Place of Exotics

The planting of exotic (introduced) tree species, both broadleaved and conifer, need not be ruled out, but the sensible approach should be to use them in situations where native species are unlikely to achieve the management objectives. Economic productivity is the key to the long-term well-being of any woodland, and exotic trees such as Sitka spruce, Japanese larch and Douglas fir can be important in woods where the production of profitable crops of timber is important. Grown in appropriate locations, such species can even contribute to landscape and amenity values. Fast-grown exotics may also be the solution where carbon fixing is important, and they may have a place in protecting the natural gene pool where native woodland is being extended. Furthermore, they can add to wildlife populations in both number and diversity while, at the same time, our wildlife adapts to the new habitats introduced species provide.[2]

Most productive forest in Britain is composed of exotic conifers, of uniform age and covering fairly large areas. But many small woodlands have also been coniferized or planted with introduced broadleaves. When conifer crops are felled, even where the plantation is on an ancient woodland site, there may be a case for replacement with another crop of conifers (although many owners choose to replace conifers with broadleaved trees, wholly or partly). If the woodland is already broadleaved, any replanting should normally be with native or traditional broadleaved species, though it may be appropriate to incorporate up to 50 per cent of compatible conifers in any replanting, provided they are managed so that the continued long-term broadleaved character of the wood is assured.

In broadleaved woodlands of more recent origin, planted or seminatural, the management aim should be to maintain the character of the area. Any broadleaved species may be appropriate for restocking, although native trees or species traditionally used in the locality are usually preferred, and up to 50 per cent of conifers may be found acceptable.

MANAGEMENT OBJECTIVES

Defining management objectives is the responsibility of the woodland owner. There can be many reasons for managing small woods and any one wood may have several uses. Deciding which is to be the main one is a vital planning decision.

Once the main objective is known, any supporting objectives can be identified and ranked in order of priority. The most popular small woodland objectives are as follows:

- ■ enhance landscape value;
- ■ improve nature conservation value;
- ■ create new wildlife habitats and enhance biodiversity;
- ■ produce marketable timber; → HARD TO TRANSPORT OFF SITE !
- ■ produce minor wood products;
- ■ regenerate the woodland;
- ■ provide recreational facilities for family and friends;
- ■ provide commercial recreational facilities;
- ■ enhance sporting value;
- ■ provide shelter for agricultural crops;
- ■ provide shelter for livestock;
- ■ provide shelter for buildings;
- ■ screen unsightly features.

Other objectives may aim to:

- ■ provide public access;
- ■ build up a capital reserve for future realization;
- ■ provide employment;
- ■ secure effective integration with agriculture;
- ■ provide a secluded environment;
- ■ provide educational facilities;
- ■ conserve archaeological features;
- ■ provide water catchment protection;
- ■ create a sound reduction barrier;
- ■ preserve the area as woodland for future generations.

WOODLAND CLASSIFICATION

With the objectives established, the next requirement is to understand the type of wood under consideration. Several systems of woodland classification are in use, some of which recognize a bewildering range of categories. These complex systems require training in their application and are limited to those with a sound grasp of woodland botany. A less sophisticated but quite adequate system for day-to-day management recognizes just four major woodland types, as detailed below.

Seminatural Ancient Woodland

These have existed since at least 1600 and are composed of locally native trees and shrubs that derive from natural growth or regrowth rather than from direct planting. Most have been cut down and regrown many times. These woods are of special value because of their long and continuous history. They are the nearest we have to the original prehistoric woodland, though the majority have been modified by centuries of management [see Appendix 1: Recognizing Ancient Woodlands].

Replanted Ancient Woodland

These have been continuously wooded since 1600 but have at some time been felled and replanted (often with conifers). Where plants and animals have survived from the former woodland, these woods may acquire some of the characteristics of seminatural woodland, particularly where the planting has been with native species.

Secondary Woodland

Secondary woods have developed naturally on land previously cleared for agriculture. They may be hundreds of years old or of relatively recent development. The older ones may harbour relic plants from prefertilizer times and can be of high environmental value, though they are not usually as valuable as seminatural ancient woodlands because of their shorter ecological history.

Plantation

These have come about from the planting of trees on land previously cleared for agriculture.

WOODLAND ASSESSMENT

A survey is essential if a woodland's current value and its potential for improvement are to be fully appreciated. The type and extent of the survey will be dictated by the objectives. For example, management primarily for timber production will call for detailed tree measurements and estimates of growth potential, whereas management mainly for wildlife conservation will usually depend on an environmental audit and an appraisal of habitat potential.

Neglected woods can be difficult to survey because parts may be virtually impenetrable, but effective planning relies on accurate intelligence. The woodland's most important features will need to be identified before improvements can be considered. The potential benefits of obtaining independent and professional advice cannot be overstated. Owners with little or no knowledge of woodland management would be well advised to avoid DIY assessments. In spite of who carries out the assessment, however, a certain amount of information will be needed if the various management options are to be properly evaluated and the most appropriate course of action taken. The information will come from several sources.

Aerial Photographs

Local authorities will often have sets of aerial photographs. These may be available for public inspection through the local planning department or the public library service. Their use can save many hours of survey work on foot. Aerial photos can show:

- species and area changes over time;
- direction of rows (where trees have been planted);
- location of unusually large trees;
- path of power lines;
- course of streams;
- layout of road and ride systems;
- potential for new access routes;
- extent of wind damaged areas;
- areas affected by disease;
- location and extent of glades and ponds;
- previously unrecorded archaeological sites.

Maps

Basic surveys call for good quality maps. Ordnance Survey (OS) maps of 1:2500 scale are ideal for small woodland work. Before the survey, the map (or an overlay) should be marked to show:

- ownership boundaries;
- rights of way;
- wayleaves and easements (above and below ground);
- scheduled archaeological sites;
- areas subject to statutory or voluntary agreements.

Survey

In the wood the surveyor will need a notebook ('all-weather' notebooks are useful) and a map. A thorough and systematic inspection should enable the plotting onto the map of such features as:

- roads (differentiate between paved and unpaved surfaces);
- paths;
- ponds, streams and rivers;
- glades and rides;
- rock outcrops;
- hollows and large canopy gaps;
- ditches;
- earthbanks.

The following crop types should be mapped as separate units:

- pure conifers;
- pure broadleaves;
- mixed crops;

Box 1.1
HOW TO ESTIMATE THE AGE OF TREES

In the absence of planting records, the most accurate way of estimating the age of a crop of trees is by felling one or two to count the annual rings. Where felling is inappropriate, a rough idea of the age of conifers can be gained by counting the whorls. Each year a new set of branches (a whorl) grows radially around the main stem. Even when branches have fallen off or have been pruned, it is usually still possible to distinguish the scars, enabling an age count to be made. While still young, some broadleaved species can also be aged by this means.

Another method involves the removal of a core of wood from the stem of a tree using a hand-held, hollow boring instrument (Pressler's borer). The number of annual rings and their distance apart will be seen on the core, allowing a rough estimate of age to be deduced. (Some experience is necessary to gain reliable estimates of age by this method.)

- special crops (for example, Christmas trees);
- non-commercial crops.

Notes should be made regarding:

- density of trees;
- main, secondary and understorey species;
- estimates of tree ages;
- tree sizes;
- tree quality;
- condition of fences, gates, bridges and drains.

In surveying woodlands that are to be managed primarily for timber production, more detailed assessments may be necessary (see Chapter 2).

MANAGEMENT OPTIONS

There can be no standard prescription for the management of woodlands. Each wood is unique, requiring its own special set of management proposals. Generally, the larger the wood, the greater the options for change. Options will vary according to the objec-

A thorough woodland assessment may reveal scarce species such as wild service trees

Box 1.2
HOW TO ESTIMATE DISTANCE WITHOUT A TAPE MEASURE

Accurate land measurement normally involves the use of ranging rods and tape measures, and for greater convenience is best done by two people. Where precision is less important, it is possible, with practice, to achieve a fair measure of accuracy simply by pacing the ground to be measured. But first, some training is required.

Peg out a 100-metre baseline in typical working conditions (for example, along a rough woodland ride). Walk along this line several times and calculate the average number of paces needed to cover the distance. For many people, the number will exceed 100. It is a mistake to try to adjust the stride to fit 100 paces into the length. When measuring, use this personal pacing factor and clock up the score using a tally counter. The factor will need to be checked for reliability from time to time.

tives. Most small woodland owners are not trying to make a profit. They accept that their woods can become commercial only to a limited extent and rehabilitation is usually for some other objective. However, few are willing or able to subsidize woodland work, and owners often impose a condition that operations, at the very least, should break even financially. By adopting a low-key, but judicious, approach to management, this can usually be achieved. Some of the more usual management options available to small wood owners are given below.

Clearance

Hardly an option – it is usually illegal to completely remove a woodland whether for agricultural or any other purpose. There have to be very strong reasons for this to be allowed.

Non-intervention

Since the majority of small woods have been unmanaged for years, there is no shortage of non-intervention areas. Nevertheless, zones of non-intervention within otherwise managed woods *can* represent positive management, though it is not often practised, even in nature reserves. Non-intervention can produce:

- conditions for the maintenance of genetic conservation;
- conditions for the study of natural processes;
- benefits to species disadvantaged by management;

- sites for the study of species that rely on shaded conditions;
- control areas against which to measure the effects of management.

The following are potentially suitable as non-intervention sites:

- long-neglected woods;
- areas unaffected by past planting;
- remote areas;
- steep-sided gulleys;
- narrow stream valleys;
- areas of storm-damaged woodland;
- parts otherwise difficult to manage.

Coppicing

Most native broadleaved trees will throw up new shoots, or coppice, from the stump after felling. Coppicing is one of the simplest and cheapest silvicultural systems (see Chapter 2). Practically all broadleaved trees can be coppiced to produce a regular supply of small roundwood on rotations of six to 30 years depending on species and potential markets. Large trees, if present, may yield sawlog material. The system is often of high conservation and sporting value.

Recoppicing may be an appropriate management option

Singling

Coppice that is well past the normal age of cutting can be grown on to produce timber-sized trees by 'singling'. This involves the removal, usually in stages over several years, of all but the best stem on each stump.

Thinning

Thinning involves the removal of selected trees in a wood to achieve some particular objective. It seeks to maximize the potential of what is already present and is often employed as the first stage in the rehabilitation of a neglected and overstocked wood. Amongst its several aims, thinning is designed to improve the quality and stability of tree crops.

Continuous Cover

This system aims to produce woodland whose overall appearance, viewed from a distance, appears to be unchanging. It has a long established history of continental practice. It involves the management of irregular crops of mixed species of uneven ages. Restocking is preferred by natural regeneration but planting is not ruled out. It can be used to restore broadleaved woodlands that have been made unproductive by past exploitation or through neglect, however, the system requires considerable skill. A loose association of enthusiasts has formed the Continuous Cover Forestry Group to promote the practice in Britain, with modification according to our physical and economic conditions. Practical workshops are held.

Enrichment

This is the restocking of gaps in the woodland canopy. Patchy and understocked woods that contain open areas with a diameter at least 1.5 times the height of the surrounding trees may be enriched either by encouraging natural regeneration or by direct planting. Planting should be concentrated in the centre of the opening to maximize the amount of light available. Areas for restocking will need adequate preparation, weed control is essential, and the sites should be inspected annually to ensure that new trees are not being overshaded.

Group Felling

The felling of groups of trees, followed by replanting or the encouragement of natural regrowth, is an effective and environmentally acceptable way to improve a neglected wood. It can be used to increase productivity and diversity, or to reduce the number of

Group felling and restocking provides a diverse woodland structure and enhances conditions for plants and wildlife

unwanted, (for example non-native) trees. The size of the groups can be varied to meet the silvicultural needs of the species to be replanted.

Clearfelling

Clearfelling involves the felling of relatively large areas or even whole woods. Small woods are not easily managed for sustained yield by group felling, but the alternative – clearfelling – is generally regarded as undesirable on environmental grounds. Sometimes, however, few other options exist; for example, very small even-aged woods may have to be clearfelled where partial removal of the crop would render the remaining trees liable to wind damage.

MANAGEMENT CONSTRAINTS

Constraints to woodland management can cover several fields. For instance, management can be rendered more difficult where a tree preservation order is imposed (landuse constraint), where salt winds blow (site constraint), or where grey squirrels abound (biotic constraint). Most problems will be overcome with a little perseverance; some may call for professional assistance; few should be sufficiently forbidding to cause plans to be abandoned.

The following sections are intended for guidance and information, and then only in the broadest sense. Inevitably, matters have had to be greatly simplified, so no statement should be taken as definitive.

Landuse Constraints

The law governing the use of land in Britain is complicated and extensive, and it frequently differs across national boundaries. Regulations that affect England may not apply in Wales or Scotland. Rights, responsibilities and arrangements with various government departments, non-government organizations and local planning authorities are seldom uniformly applied. For example, the Countryside Commission's responsibilities now apply only to England. The law governing the use of the countryside is the same in Wales as in England – but a separate body, the Countryside Council for Wales, has been established. It is that body which should be contacted for advice or information relating to Wales. Scotland has a different legal system, so what is read here need not apply to the Scottish countryside. For information on rights and responsibilities in Scotland, the reader should contact Scottish National Heritage.

National Parks

National Parks comprise extensive tracts of open countryside that afford opportunity for recreation. They were chosen for their natural beauty and easy access from centres of population and are administered by local planning authorities. In contrast to parks or reserves in many other countries, all land in a National Park continues to be owned, often privately. It may be farmed, forested, used for shooting or employed in some other way. The normal rules about access to land still apply. There is no general right of public access simply because the land is within a National Park.

No work is allowed in a National Park that will substantially reduce access (except for short periods when advance notice must be given). In general, tree work is more tightly controlled than elsewhere. The intention of this is to maintain the character of the area which could easily be damaged by poorly designed planting and felling schemes.

Areas of Outstanding Natural Beauty (AONB)

These are designated by the Countryside Commission. They are areas of outstanding national landscape importance. The boundaries of AONBs are clearly marked on OS 1:10,000 maps and are always shown on structure plans and local plans for an area. These plans will usually have forestry provisions and policies within them. They are generally available in local reference libraries and at local authority planning offices.

Applications for tree felling licences (see Chapter 2) or planting grant schemes in AONBs will be subject to close scrutiny, particularly with regard to landscape design of felling coupes and choice of species for planting or replanting.

Sites of Special Scientific Interest (SSSI)

These are designated by English Nature, the Countryside Council for Wales and the Nature Conservancy Council for Scotland. They are areas of land that contain features of local or national rarity. There are also 'second tier' sites of conservation importance. These may be identified in structure plans or local plans and local authorities may have details.

An owner whose land is designated a SSSI (an ASSI in Northern Ireland) is sent a notification and a list of potentially damaging operations that should not be undertaken without prior consultation with the relevant agency. Under the terms of the Wildlife and Countryside Act 1981, payments may be made, in certain circumstances, for revenue forgone in SSSIs in the interests of nature conservation.

National Nature Reserves (NNR)

English Nature, the Countryside Council for Wales and the Nature Conservancy Council for Scotland have a statutory obligation to establish, maintain and manage nature reserves. Some of the land in NNRs is owned or leased by the agencies. The rest is owned by other interests so the agencies must work with private landowners and occupiers through management agreements. These agreements may contain restrictions on the work that may be carried out on the land, but there may be compensation for revenue forgone.

Local Nature Reserves (LNR)

Local authorities have the power to create nature reserves within their own area of control, but this is not a designation that can be imposed on an owner. LNRs are created by agreement.

Environmentally Sensitive Areas (ESA)

These are farming areas designated for their national ecological or landscape importance which are under threat from actual or potential changes in farming practice – including the effect of farming on woodlands.

In designated areas, financial assistance is available to farmers if they agree to follow special requirements. The details of each ESA scheme vary and payments differ. In some cases special payments may be available to enhance the conservation value of existing farm woodland. Farmers with ESA payments still have access to woodland grants, providing there is neither overlap of finance nor conflict of aims between the schemes. [Further details are available from regional offices of the appropriate agriculture department.]

Archaeological Features

Woodlands may contain archaeological features that are 'scheduled' under the terms of the Ancient Monuments and Archaeological Areas Act 1979 (AMAAA).[3] These represent nationally important monuments and any management that will result in their damage requires consent. However, some activities do not require consent and these may include forestry works, provided that the activities formed the regular landuse in the previous five years. It does not extend to drainage, ploughing, tree planting, fencing and the grubbing out of trees.

The act allows grants for management agreements on monuments (whether scheduled or unscheduled) relating to ongoing maintenance and management, including scrub management, pest control and fencing. Under the AMAAA, landowners must protect any scheduled ancient monument, field monument or ancient building on their property. Where archaeological features do occur, it may be possible to incorporate them into glades or rides. The Forestry Authority will not approve a grant scheme application that does not provide for the protection of important archaeological sites, including non-scheduled sites. The onus is placed on owners to seek information about the archaeology of their woodland. Not all sites of archaeological interest are marked on OS maps. There are large areas of the country where information about archaeological features is scanty.

Some areas of woodland are themselves archaeological features within the landscape.[4] In some instances, a piece of woodland with its distinctive shape may have remained unaltered for centuries.

It is an offence to use a metal detector on a scheduled ancient monument or area of archaeological importance to detect or locate objects of archaeological or historical interest without written consent.

Woods Adjacent to Designated Sites

Management in woodlands adjacent to SSSIs, NNRs, AONBs and National Parks may, in some circumstances, affect the interest of the designated area. Proposals may, therefore, attract objections, even though they are not on the site itself.

Public Rights of Way (RoW)

The main responsibility for administering and enforcing the Rights of Way Act 1990 falls on the highway authority (the county, metropolitan district or outer London borough council).[5] The highway authority has a statutory duty to:

- protect and maintain RoWs;
- prevent the obstruction of a RoW;
- ensure that land is properly restored after disturbance;
- ensure that growing crops do not inconvenience the use of a RoW;
- signpost RoWs where they meet a metalled highway;
- provide additional signs and waymarks where they are necessary;
- prepare and keep up to date a 'definitive map and statement'.

If a RoW is shown on the definitive map, this is conclusive evidence, in law, that the public had those rights – and have them still unless there is a legally binding change. However, the reverse is not true; it is still possible for a RoW to exist and to have been omitted from the definitive map, or for the status of the path to be incorrect. The OS 1:25,000 scale Pathfinder map series shows public RoWs in green, though older Pathfinder maps may not show them accurately. For some areas of attractive countryside, 1:25,000 scale Outdoor Leisure maps have been produced that show additional information, such as waymarked routes, camp sites, permissive paths and some areas of open access.

There are four types of rights of way:

- footpath: may be used on foot only;
- bridleway: may be used on foot, horse and bicycle;
- byway open to all traffic (BOAT): may be used on foot, horse, bicycle and vehicle;
- road used as a public path (RUPP): may be used on foot, horseback and bicycle, unless it can be proved there are other rights – these are subject to reclassification (a long and sometimes controversial process, since determining exactly what rights exist can be difficult).

An invalid carriage, pram, pushchair or wheelchair may be used along any RoW, and dogs are permitted providing they are on a lead or under close control.

Landowners and occupiers have a responsibility to respect the public's right of passage. They should do nothing that will inconvenience or endanger the public in any way. It is an offence to erect a new fence across a public highway without the permission of the highway authority, and it is generally illegal to plant tree crops on a RoW. Owners should check the condition of trees on land adjacent to ROWs to ensure they pose no danger to the public. When trees near to RoWs are being felled, owners must always be aware of their responsibilities under current health and safety at work legislation.

Nearly all RoWs are 'maintainable at public expense'. The surface is considered the responsibility of the highway authority, but overhanging vegetation should be in the care of the landowner or occupier. The owner or occupier must also keep any gates or stiles in good repair and easy to use. In some circumstances, grants may be available from the local authority for repair or renewal of gates and stiles.

Permissive Rights of Way

Given adequate controls, owners may choose to welcome the public into their woods under access agreements with the local authority. Permissive Rights of Way should not, therefore, be a constraint on woodland management. Nevertheless, owners need to be careful that rights are not created accidentally by allowing arrangements to overrun the agreement period. A landowner can ensure that permitted access does not lead to any statutory declaration of a Right of Way in England, Scotland or Wales by putting up signs stating 'Permissive Path Only' and by closing the path one day a year and erecting a sign stating 'Closure to Avoid Dedication' (Christmas Day is often chosen for this purpose). Under a further provision in England and Wales a landlord can deposit a map with the highway authority showing the paths that have been accepted as public, together with a statutory declaration that no other paths are intended. This gives protection and must be renewed every five years. Permissive Rights of Way may be eligible for some financial assistance.

Protected Access

This is where access is granted to specific individuals or groups (for example, school parties) by invitation or request. Access may be restricted to permit-holders or key-holders, or may involve a more formal agreement. Access may also be granted to the general public for a limited period during the year (for example, over a bank holiday).

Permitted Access

This can be the owner's informal toleration of people entering the woods as and when they please, or it may involve some measure of control that encourages people to keep within certain areas.

Occupiers' Liability

Woodland owners have a duty under the Occupiers' Liability Act 1967 'to take such care as is reasonable to see that the visitor is reasonably safe in using the premises'. In mature woods this may involve regular inspection of formal routes for hazards, such as dangerous trees or potholes in surfaced paths. Other precautions may include fencing off steep drops or old quarries and using signposts to warn visitors of potential dangers.

Tree Preservation Orders (TPO)

These are designated under the Town and Country Planning Regulations.[6] They are made by local planning authorities to protect selected trees or woodlands if their removal would have a significant impact on the environment and its enjoyment by the public. Trees subject to a TPO should normally be visible from a public place such as a road or footpath. Details of confirmed and provisional orders are kept at the local authority offices.

In general, a TPO makes it an offence to cut down, top, lop, uproot, wilfully damage or wilfully destroy a tree without the planning authority's consent. Penalties can be severe – in 1995 a builder was fined £25,000 for the 'wilful destruction' of two sycamore trees that were subject to a TPO. He was also required to pay costs.

A landowner whose trees become subject to a TPO must be informed by the local planning authority making that order. Similarily, the seller of land is required to notify the purchaser that a TPO has been made in respect of trees on the property. If the Forestry Authority has given aid under a grant scheme, a TPO can only be made if there is no current plan of operations, and only then with the Forestry Authority's approval. Trees subject to a TPO remain the responsibility of the owner.

Conservation Areas

These are areas of special architectural or historical interest, the character or appearance of which it is desirable to preserve or enhance.[7] They are designated by local planning authorities and are often, but not always, centred around listed buildings. Other buildings and features, including trees, may also contribute to the special character of a conservation area. The usual provisions apply to trees protected by tree preservation orders. For others, the local planning authority must be given six weeks' notice in writing if work is to be carried out on them. No work should be carried out during that period without permission. Anyone who does could be fined and may be required to plant another tree. However, permission is not required to cut down or work on trees less than 7.5 centimetres (cm) in diameter, measured 1.5 metres (m) above the ground, or 10 cm if this is to help the growth of other trees. Other exceptions also apply. Anyone in doubt should check with the local authority.

Easements and Wayleaves

The presence of overhead power lines and underground services (such as electricity, gas, water or drains) can influence woodland management. Any work or tree planting in their vicinity may require the prior permission of the appropriate authority.

Land Tenure

The often complex relationship that exists between ownership, tenancy and woodland management can pose problems.[8] Most leases, as opposed to life tenancies, reserve all trees to the landlord, with access for maintenance, felling and extraction. This divorces the management of the land from the management of the trees. The well-being of the woodlands, therefore, may be of no consequence to the tenant.[9]

Multiple Ownership

Woods in which several people own different parts, and woods that are jointly owned by several people, can also make for difficult management.

Site Constraints

Each species has characteristics that suit it to its own natural habitat – though occasionally an exotic species can outperform native species (for example, the exceptional growth of Sitka spruce in Britain). In general, the climate, topography and soils of Britain provide ideal conditions for tree growth, but individual site factors are nevertheless important and there are limits beyond which certain species will not thrive. Ensuring that woods are composed of the most suitable species to match the site and to meet the owner's objectives is crucial (see Chapter 3).

Temperature

Temperature has an important influence on tree growth and can limit the use of some species. For example:

- Early frosts in autumn may damage shoots that have not hardened off.
- Late frosts in spring can damage tender shoots that have just started to grow.
- In summer the sun's heat can scorch the foliage of seedling trees and can occasionally cause older trees to wilt and die back.

Light

The amount of light available to trees governs the degree of photosynthesis. Most trees grow better in full light conditions than in shade, though there are exceptions.

Rainfall

Different tree species have different water requirements. Long summer droughts can cause damage to trees, such as beech, that are sensitive to water stress. In low rainfall areas, some species – spruces, for example – will only grow well on moist soils.

Snow and Ice

Clinging wet snow and ice formation can cause branches to break. The damage is seldom critical in England but can be locally severe in Scotland and Wales.

Wind

Wind behaviour is dependent upon topography, altitude, aspect and degree of slope. Trees become increasingly susceptible to damage by wind as they grow taller. In lowland broadleaved woods the problem is seldom serious. Sites are generally no more than moderately exposed and soils are usually of sufficient depth for good root development. On the windiest sites damage is usually limited to crown deformation, desiccation and poor growth. Conifer woods, in contrast, are often planted at higher elevations and on shallower soils where windblow (when whole trees are uprooted) and windbreak (when main stems are snapped) can pose serious problems. Damage can be rated as sporadic (when individual trees or small groups are damaged) to catastrophic (when whole woodlands are brought down over a wide area).

Topography

Topography and climate are inextricably linked. For example, frost damage is liable to occur on clear, still nights when cold air flows downslope to collect in valleys and hollows.

Altitude

In general, temperature and the length of the growing season decrease, while exposure increases, with altitude. Few woods are economically productive above approximately 300 metres.

Slope

Some of the steepest slopes can support good tree growth but management may be impossible due to the limitations of movement by man and machine.

Aspect

Aspect seldom influences tree growth to any significant degree in lowland Britain – though walnut generally grows best on a southerly aspect. At higher elevations there can be a marked contrast between the vegetation of north- and south-facing slopes at the same elevation.

Soils

Soils have a major effect on tree growth. They need to be sufficiently deep to allow tree roots to effect strong anchorage and to provide adequate nutrition. Some conifers, such as spruce, have superficial rooting systems and are able to prosper on relatively shallow soils, but most broadleaved trees require deeper soils if they are to grow well.

Soils vary in their fertility. Most lowland woods occupy sites at the more fertile and desirable end of the range and will support good growth in a wide variety of tree species. Soils of upland moors, sandy soils and those overlying chalk and limestone are often less favourable for tree growth, and wet soils can pose special problems. Compacted soils, whether created by woodland operations or of natural origin, create difficulties for tree root development. Some upland soils contain an ironpan that can form a barrier to roots, while former arable land may contain a plough-pan that can have the same effect.

The condition known as lime-induced chlorosis can occur in immature trees on chalk and limestone soils. Conifers are particularly prone, exhibiting pale, yellowing foliage and increasingly poor growth. Further information on the suitability of various tree species on a range of woodland soil types can be found in Chapter 3.

Biotic Constraints

The incidence, or likely incidence, of certain woodland pests and diseases can influence planning decisions. The damage that some of the most common agents can inflict on trees and woods is considered here. Control is covered later.

Ants

Ants can be a problem on old meadows where treeshelters are in use. They build their nests within the shelters and this can result in tree deaths.

Weevils and Beetles

Pine weevils breed in the stumps and roots of a wide range of conifers.[10] They feed on the living bark and roots of most woody plants. Felling a coniferous crop produces a large increase in breeding material, while material suitable for feeding is reduced. Young trees used for restocking are liable to be heavily attacked.

Small Mammals

Woodland mice, bank voles and field voles eat tree seed and strip the bark of roots and the lower stems of young trees. Where stems are completely girdled this can result in death. Damage tends to be localized and is most likely in years when small mammal numbers are high.

Grey Squirrels

Grey squirrels are a major pest. They ruin horticultural crops, rob birds' nests of their eggs, kill nestling songbirds, eat buds and seeds, and cause significant damage to trees by stripping bark, anywhere from the base to the topmost branches.

Rabbits

Rabbits can cause severe damage to woodlands. They eat young seedlings, cut off leading shoots, browse branches and gnaw bark.

Hares

Hares can be almost as destructive as rabbits, though they are far less numerous.

Deer

Deer are widespread in the British countryside. They browse on tree leaves, shoots and buds (particularly coppice), strip bark for

The grey squirrel needs to be controlled if broadleaves are to be successfully grown

food and fray tree stems to mark territories and to clean antlers by rubbing off the velvet. At the same time, they can create small clearings and gaps in which light-demanding species will survive.

Cattle

Cattle sometimes have access to woods, either as a result of deliberate farm policy or because fences and hedges are no longer maintained. Cattle graze on most plants, including woody plants. They also trample small trees into the ground, thus frustrating all hope of a wood regenerating naturally. By the action of hooves on wet sites they can poach the soil, causing serious damage to mature woods and ruining young plantations while severely compacting waterlogged soils.

Sheep

Sheep can be a major problem in upland woods. They graze on selected plants and browse on most trees. Most damage, including bark stripping, occurs in bad weather when sheep move into woods for shelter.[11]

Other Livestock

Horses, pet ponies, donkeys and goats will eat tree foliage and gnaw bark.

Dutch Elm Disease

Most of Britain is now affected by Dutch elm disease. In the majority of cases this has ruled out the use of elms in planting and replanting schemes.

Honey Fungus

Honey fungus is found in many woodlands. It can kill trees as individuals or in very small groups. It rarely causes extensive damage, though it may rule out the use of particularly susceptible species.

Fomes Butt Rot

Fomes, a fungal root and butt rot disease, is generally only a major problem where conifer trees are grown on a commercial scale.

ADMINISTRATION

With the objectives established, the woodland surveyed, the tree crops assessed, the guidelines considered, the options defined and any constraints understood, a plan of action can be drawn up.

The Management Plan

This is the basic document, usually accompanied by a map, that sets out in detail the proposed management decisions. The amount of detail required will depend on the size and complexity of the wood. A typical plan would normally comprise a:

- financial appraisal;
- plan of operations;
- compartment schedule;
- list of requirements (labour, materials and machinery);
- set of maps;
- review date (to ensure continuity and flexibility of management).

Any financial appraisal should take account of:

- the costs of various management options;
- any professional fees;
- ongoing management costs;
- ongoing maintenance costs;
- the availability of grant aid;
- any likely revenues from charges, lets, etc;
- any likely income from sales.

If desired, this information can be used to produce a projected cash flow to ascertain levels of commitment at any given time and to assess the viability of the scheme in the long term.

For most small woods, however, such detail will be unnecessary. A simple five-year plan of operations will usually be adequate. This should provide:

- ownership details;
- property details;
- the objectives;
- the long-term aims;
- a brief site description;
- notes on any features of special interest;
- a schedule of work proposals;
- an indication of when tasks will be undertaken;
- a statement of areas involved;
- an annotated map.

An example of a plan of operations for a neglected small wood will be found in Appendix 2.

Grant Aid

The availability of grant aid is usually an important consideration but information on grants soon gets out of date. There are many sources, changes occur frequently and new grants are introduced from time to time. Consequently, this section deals only with the main sources of grant aid and then only in the broadest terms.

Enquiries regarding grant aid for tree and woodland work are usually best directed, in the first instance, to the local authority. Intending applicants should then contact the organization concerned and ask for the latest details. The need to ensure that public money is well spent and that plans are acceptable in silvicultural and environmental terms means that applications will be

Box 1.3
HOW TO PLOT INFORMATION ON A MAP

A map of 1:2500 scale is usual for recording and displaying information. Woodland boundaries should be marked with a solid line no more than 0.5 millimetres (mm) wide. The inside of the line should show the external boundary of the area. Use a line thus —·—·—·— to show any internal compartment boundaries and number the compartments (1,2,3, etc). If further division is appropriate use a line thus ---------- and add subcompartment letters (1a, 1b, 1c, etc). It is customary to amalgamate similar crop types into management units by colour coding or crosshatching. The map should clearly show the scale and the map reference of the centre of the wood and main access point. Remember that OS maps are copyright.[12]

Box 1.4
HOW TO MEASURE MAP AREAS

Accurate map area measurement calls for professional equipment, but a reasonable estimate can be obtained with the aid of a simple perspex 'hectare grid' of appropriate scale. The grid is placed in a random position over the area to be measured and the subdivisions counted. The average of three counts is usual. The edge of the grid is demarcated for the measurement of distance. Area grids are generally available from OS map agencies.

considered very carefully by the grant aiding authority. In some cases that body will need to consult with other interested organizations. Owners must allow for the fact that many weeks may elapse before an application is finally approved. Applicants should:

■ provide good quality maps;
■ keep copies of all submissions;
■ note dates of application submissions and returns;
■ never carry out work until plans are approved.

Woodland Grant Scheme (WGS)

It is through the WGS that the Forestry Authority currently gives grants to create new woodlands and to manage existing ones. Owners and their agents can apply to their nearest Forestry Authority Conservancy Office for a WGS applicant's pack.

Applicants are required to manage their woodlands in accor-

Box 1.5
HOW TO READ A MAP REFERENCE

Six-figure map reference numbers are accurate to the nearest 100 metres. They are useful for identifying the approximate position of the centre of a wood or an access point.

A common mistake is to get the figures the wrong way around. Remember the rule: 'along the hall and up the stairs'. Start in the south-west corner of the map and move across the baseline until vertically below the point to be sampled. Note the two large digits of the gridline to the left and then add a third digit by estimating the number of tenths eastward. If the reference to be read sits squarely on the gridline then the third digit is 0. Repeat the process using the numbers up the side of the map.

---

dance with an approved plan. Where an application involves seminatural ancient woodland, the Forestry Authority will compare management proposals with the advice contained in their advisory guides. Applicants are free to propose other forms of management for these woods, but must satisfy the Forestry Authority that their proposals will be effective in maintaining, and preferably enhancing, the special characteristics of the woodland. The advice given in these guides is intended to create a flexible framework rather than a straitjacket, so that woods and their owners can develop their individuality as much as possible without reducing options for future generations.

There are disincentives to applying for the scheme. The rules are complex, application forms are confusing to novices, and maps must be meticulously drafted. Owners wishing to plant a couple of hectares must supply information similar to that required for a large forestry estate. The Forestry Authority should always be mindful that initial enthusiasm is easily dissipated by excessive bureaucracy.

Farm Woodland Premium Scheme (FWPS)

The FWPS offers annual payments over and above WGS grants to farmers and crofters converting agricultural land to woodland, subject to certain conditions. The payments are to compensate for agricultural income forgone. It is not possible to apply for the FWPS alone because the environmental and silvicultural standards of the WGS must be satisfied before a FWPS application can be approved. Only one joint application is needed. Information on the FWPS will be found in the WGS applicants pack.

The Countryside Council for Wales and Scottish Natural Heritage may offer financial assistance towards planting and managing woodlands for landscape objectives. Commonly administered by county, district or borough councils, these grants are normally up to 50 per cent of total eligible costs. Proposals must show that there will be clear public benefit. Further details are available from the relevant local authority.

English Nature, the Countryside Council for Wales and Scottish Natural Heritage may give grants towards planting and managing woodlands where nature conservation is the main objective. National Park authorities may give financial assistance for woodland management within National Parks.

The Need for Felling Licences

A licence from the Forestry Authority is normally necessary to fell growing trees (though not for topping or lopping), but in any calendar quarter, owners or tenants can fell up to five cubic metres without a licence as long as no more than two cubic metres are sold. A licence from the Forestry Authority is not needed where:

- The felling is included in an approved plan under a Forestry Authority grant scheme.
- The trees are growing in a garden, churchyard or public open space.
- When measured 1.3 m above the ground the trees are: all less than 8 centimetres in diameter; if thinnings, less than 10 centimetres in diameter; if coppice or underwood, less than 15 centimetres in diameter.
- The trees interfere with development permitted by planning legislation or work legally carried out by public organizations.
- The trees are dead, dangerous, causing a nuisance or are badly affected by Dutch elm disease.
- The felling is done under an Act of Parliament.

Those who apply for a licence should be:

- the owner of the land on which the trees are growing;
- a tenant where the lease entitles the individual to fell the trees;
- an agent acting for the owner or tenant.

Timber merchants and contractors must make sure that a licence has been issued before they carry out any felling, since they could be prosecuted if there is no licence. Applications cannot be accepted from timber merchants or contractors unless they come into one of the above categories. Applications call for accurate area measurements, a tally of trees and a reasonable estimate of timber volume.

To allow time for the woodland officer to look at the site and carry out any necessary consultations, the application with a map of the area should be sent in at least three months before the proposed date of felling. Where the Forestry Authority considers it necessary, replanting conditions will be added to the licence, in which case grants may be available.

No felling should take place until a licence has been issued. Any felling carried out without a licence is an offence, unless it is covered by one of the exceptions (more fully set out in the Forestry Act 1967 and related regulations). This can mean a possible fine of up to £2000 or twice the value of the trees, whichever is the higher.

The Forestry Authority also has the power to serve a notice on an owner or tenant convicted of an illegal felling to restock the land concerned, or any other land as may be agreed. If the conditions of a felling licence or restocking order are not met, the Forestry Authority may issue an enforcement notice demanding that action be taken to meet the conditions. In the event of failure to carry out the action named in an enforcement notice, the Forestry Authority can go on the land, carry out the work and recover the expenses from the person served with the notice. It is an offence not to obey an enforcement notice and can mean a possible fine of £2000.

Other Permissions

In certain circumstances, whether or not a licence is needed, special permission may be necessary from another organization for any proposed felling. This can apply where the trees are in a Conservation Area or a Site of Special Scientific Interest (SSSI) or are covered by a Tree Preservation Order (TPO).

Insurance

A number of insurance companies have schemes designed specifically with the needs of the forestry and woodland industry in mind. Some offer generous discounts for members of certain professional associations. Cover is generally available for the following contingencies.

Windblow

Many woodland owners have suffered losses as a result of gales. In some cases revenue from the sale of damaged timber has not covered the costs of clearing and replanting the site. Insurance can protect the woodland owner against this situation and can also provide cover for the loss of expectation associated with the loss of a crop before maturity. Normally, crops up to 50 years of age can be insured against damage by windblow. Owners of low-risk woodlands can usually opt to bear an increased excess in consideration of a lower premium.

Fire

This will normally cover damage caused by fire, lightning, aircraft, explosion and earthquake. Cover can also extend to fire-fighting costs.

Property Owner's Liability

This cover will safeguard woodland owners against claims from the general public. The owner can usually select the level of cover required and the level of risk to be insured.

Employer's Liability

This covers legal liability for damages and costs for injury to employees arising out of or in the course of their employment. There is a legal requirement for this insurance for all direct employees, for labour-only subcontractors or self-employed persons.

Public and Products Liability

This covers legal liability to third parties (non-employees) for either injury or damage to property as a result of negligence arising out of work done or goods supplied in connection with a woodland business. The policy will usually cover all woodland activities including the use of herbicides, pesticides and explosives, but may not cover burning of debris warranty.

Equipment

This covers loss or damage to plant and equipment.

Professional Indemnity

This is essential for anyone offering professional advice. It covers legal liability to third parties for breach of professional care caused by neglect, error or omission.

ORGANIZING THE WORK

The question of whether the work set out in the management plan will be undertaken by the woodland owner or a contractor, or will be put in the hands of a professional management consultant, is likely to turn on such factors as the:

- objectives of management;
- owner's resources;
- owner's level of expertise;
- level of commitment to future work;
- assumed benefits.

The Woodland Owner

Owners have the option of carrying out some or all of the work needed using their own labour and equipment. They may choose to do this to save the cost of employing skilled assistance, because they consider the job too small to employ a contractor or to get the benefit of some healthy exercise. Owners will know their woods better than others and are well placed to understand the terrain, ease of access or likely constraints. It is important to remember, however, that in inexpert hands, forestry tools (and trees) can be lethal. In addition to cuts from chainsaws, there are perils associated with edged tools, falling trees, power lines, harvesting and extraction machinery and stacked produce. Farm tractors can be used for transporting plants, fencing materials and equipment, and for extracting light produce such as firewood – but only tractors designed for the job or suitably modified should be used to extract heavy logs. Accidents happen because people fail to use appropriate safe working practices.

Owners who choose to undertake work in their own woods should consider the following advice:

- Chainsaws should be used only after training.
- No one should work alone in the woods, even with hand tools.
- Safety footwear and non-snag clothing should be worn for all work.[13]
- A safety helmet is essential whenever there is a risk of head injury.
- Safety helmets need to be replaced if damaged (otherwise at intervals as advised by the manufacturer).

- Tree climbing, delimbing and the felling of large trees should be left to a contractor.
- When working with noisy equipment, hearing protection is required.
- In many cases eye protection will also be needed.

A selection of hand tools and other items will be required. A basic kit might comprise:

- bowsaw;
- pruning saw;
- axe;
- sharpening stone;
- files;
- slasher;
- billhook;
- reap hook;
- spade;
- first aid kit.

For the more ambitious DIY owner, a few items of powered equipment are desirable:

- chainsaw;
- brushcutter;
- grass mower;
- woodchipper;
- tractor (with power take off);
- circular sawbench;
- 4 x 4 vehicle.

When using a chainsaw the following items should be worn:

- safety helmet;
- ear protection;
- eye protection;
- upper body clothing that is close fitting;
- gloves with protective guarding on the back of the left hand;
- appropriate leg protection;
- chainsaw operator's boots (incorporating protection for the toes, top of the foot and front of the lower leg; they should also have a good grip).

Brushcutter and clearing saw operators should wear:

- safety helmet;
- ear defenders;
- suitable eye protection;
- thorn-proof gloves;
- safety boots;
- robust outer working clothing to suit conditions;
- harness for supporting the machine.

When using herbicides and pesticides, chemical-proof protective clothing, no less than that specified on the product label, should be worn. This might include:

- face shield or particle mask;
- coverall;
- hood;
- gloves;
- boots.

Protective clothing can be hot and uncomfortable. If heat generation exceeds heat loss, work must stop to allow the body to cool off.[14] Obviously, standards for personal protective equipment are subject to change. Guidance is given in leaflets published by the Health and Safety Executive and the Forestry and Arboricultural Safety and Training Council (FASTCo). Spraying should only be undertaken after training.

Other Health Matters

Vibration White Finger

Chainsaws and similar hand-held forestry equipment have been cited as possibly causing operator problems.[15] Vibration can cause damage to the arteries, nerves, bones and muscles leading to a potentially painful condition known as hand-arm vibration syndrome. The most common result of injury is vibration white finger. These tools do not always cause injuries. Much depends on the type, how they are maintained, the way they are used and how long operators work with them.

Lyme Disease

Lyme disease is caused by a bacterium that is transferred from its wildlife host (mainly deer, but also cattle and goats) to man by ticks.[16] In the mildest form, symptoms may be no more than a slight fever, malaise and a rash, often in a circular or bulls-eye form. When diagnosed early, with appropriate antibiotics a rapid cure is obtained, but in more serious cases the nervous and cardiac systems may become permanently affected. A doctor should be consulted immediately if a rash or other symptoms associated with Lyme disease develop. Sensible precautions should be taken in tick-infested areas, including the wearing of long-sleeved shirts and long trousers, the use of repellents and brushing off clothing before going indoors.

Tetanus

Tetanus is a serious life-threatening disease caused by a micro-organism that lives in soil. It gains entry to the human body by way of deep, penetrating wounds. Those who work in woods and forests should obtain protection by periodic injections of tetanus vaccine.

Employing Contractors

Many small wood owners will choose to use contractors as an alternative to doing the work themselves. A competent contractor will have skilled labour, adequate and properly maintained equipment, and a good working knowledge of timber specifications, prices and markets. However, to maintain some degree of control the owner should have a fairly precise idea of what is to be achieved – hence the need for a management plan.

Contractors who undertake woodland work include a few major forestry management companies as well as a large number of smaller organizations, partnerships and individuals who operate on a regional or local basis. In obtaining advice that is charged for, the owner should make sure that:

- references are obtained;
- the work is clearly specified;
- a written quotation is given;
- there are no hidden charges;
- a written agreement is obtained.

Timber Merchants

Some contractors will sell their timber to timber merchants, others operate as timber merchants in their own right and will sell directly to the sawmill. Woodland owners are sometimes wary of approaching timber merchants. Stories circulate of resourceful characters buying trees at firewood prices in the morning only to sell them as veneer logs in the afternoon. Such tales are seldom substantiated. An altogether different picture was revealed in a study by Plimsoll Publishing.[17] It showed that 38 per cent of timber merchants were in some form of financial difficulty in 1996, and this had deteriorated from 36 per cent in the previous year.

The Forestry Contracting Association (FCA)

The FCA represents the largest forestry contracting network in the UK with contractors specializing in all aspects of woodland establishment and harvesting. All member contractors work to a code of practice and offer the assurance of a certified workforce, a recognized standard of work, appropriate insurance cover and the back-up of the association.

The Association of Professional Foresters (APF)

The APF, an association for those who derive their livelihood from forestry, publishes a directory of members offering various goods and services. It provides details of those who have achieved and are maintaining standards expected by the APF.

The Timber Growers Association (TGA)

The TGA is an association for all those involved with woodlands whether for business or for pleasure. Amongst other benefits it publishes a handbook that includes sections on timber markets, timber prices, grant schemes, woodland taxation, education and training, and a herbicide guide. There is also a section on suppliers and services, as well as a list of timber users, merchants and agents.

The National Small Woods Association

The NSWA maintains a list of practitioners. The aim of the list is to put woodland owners in contact with practitioners who can provide advice and contracting services.

Engaging a Consultant

Woodland owners unable to undertake the work themselves or to supervise the work of a contractor, can choose instead to engage the services of a professional forestry company or an independent woodland management consultant. These people can offer a comprehensive and professional range of services, including woodland assessment, design, landscaping, conservation advice, marking of thinnings and fellings, timber valuations, sales on behalf of the client, and other work.

The Institute of Chartered Foresters (ICF)

The ICF maintains a list of its members in consultancy practice that is updated annually. The members' designation, business practice, declared fields of interest and regions of the UK in which they operate are listed. The scale of professional fees for forestry work as recommended by the institute is included in the directory.

The Forestry Commission (FC)

The FC is the government department responsible for forestry in Great Britain. Established in 1919, it has been through many fundamental changes in its history, but there have been two constant threads, namely: to create forest cover in a relatively unwooded country and to create a supply of home-grown timber.

In the last few years the commission has been subject to a succession of reviews. While resources have been getting tighter, demands on the FC to widen its activities beyond the core business of growing timber have been getting ever greater. It would be unrealistic to believe that we have seen the last of these changes. Today the FC operates in four different arenas:

- As a government department it advises ministers.
- The Forest Research Agency provides research, development, surveys and related services to the forest industry, and authoritative advice in support of the development and implementation of the Government's forestry policies.
- The Forest Enterprise manages and operates the national forest estate.
- The Forestry Authority sets standards for the forestry industry, runs grant schemes to help private woodland owners, and ensures regulations for plant health and tree felling are complied with.

The research division's technical branch has undertaken much useful investigation into areas of small wood management interest.

Education and Training

Appropriate training should always be obtained before unfamiliar techniques are used, and some basic training in emergency first aid is advisable. Technical training in woodland operations is available from a variety of sources.

Agricultural colleges

Over 60 agricultural and horticultural colleges now offer part-time, as well as full-time, forestry courses. Information about a particular college is gained through its own prospectus.

Forestry and Arboricultural Safety and Training Council

FASTCo is recognized as the lead body responsible for determining competence standards in all sectors and at all levels of forestry (excluding post-harvesting industrial processing). FASTCo's Register of Approved Instructors is a UK-wide training provision whose members offer a comprehensive range of courses in forestry and arboricultural skills. FASTCo is also recognized by the Health and Safety Executive as an authoritative source of information and guidance on health and safety issues relevant to forestry and arboriculture. It publishes a series of safety guides covering major forestry and arboricultural operations.

The National Small Woods Association

This organization runs a series of national courses on woodland management. The courses are of value to small woodland owners, managers, contractors, land agents, conservation groups and others who need to know more about the potential of small woodlands and how to set about managing and marketing the woodland produce. Programmes are implemented regionally, usually in partnership with a local woodland project. On completion of each module a certificate of attendance is awarded.

Landbase Training Services

This is a national organization offering training and development programmes to all types of business. Woodland courses include:

- tree recognition;
- value of woodland to landowners;
- introduction to timber measurement;
- marketing woodland produce;
- harvesting of wood using existing resources;
- thinning woodlands;
- managing neglected woodlands;
- planning, planting and maintaining new woodlands.

There are courses on aspects of conservation, landscaping, access and health and safety, and there are also programmes for supervisors.

The Health and Safety Executive (HSE)

The HSE offers a wide range of literature relating to health and safety issues.

For further information regarding technical training it is worth searching the Training Access Points (TAP) and Educational Counselling and Credit Transfer Information Service (ECCTIS) databases – available through public libraries. There is also a *Directory of Courses in Land Based Industry* that is published by Farming Press Books.

2

Timber Production and Wood Products

Managing woods for timber production should not be seen as an alternative to managing for other purposes but as one of a number of integrated and complementary objectives. Few small woodland owners are interested in timber production alone. Their incentive is more often some form of multipurpose management, with nature conservation or landscape enhancement a high priority. They may be interested, though, to the extent that they want their woods to be self-sustaining and for operations to break even financially.

Successful timber production is based upon good planning, good organization and control, and a high level of commitment. Failure to carry out operations consistently and promptly, and to adequately protect trees, will invariably result in loss of revenue.

TERMS AND CONVENTIONS

In the timber trade conifers and broadleaves are known as softwoods and hardwoods (even though the wood of yew, a conifer, is extremely hard while that of birch, a broadleaved tree, is relatively soft; see Glossary).

Contractors and buyers commonly use a traditional form of timber measurement known as Hoppus's measure. Tree volume will be estimated as so many Hoppus feet, cubic feet, or simply 'cube'. The Forestry Authority, on the other hand, uses the metric system. Estimates of volume for felling licence applications and grant schemes must be in cubic metres. For conversion purposes:

- 1 Hoppus foot = 0.036 cubic metres;
- 1 cubic metre = 27.74 Hoppus feet.

ASSESSING TIMBER POTENTIAL

Good planning depends on a competent and thorough assessment of the woodland and its growing stock. The level of precision required will determine the type of measuring and surveying equipment needed. Amongst a forester's accoutrements might be found:

- girthing tapes;
- timber sword;
- clinometer;
- angle gauge;
- relascope;
- tally counter;
- Pressler borer;
- timber gouge;
- aerosol marking paint;
- timber measuring book.

Also essential are certain forest management tables. The Forestry Commission, mainly with the professional in mind, publishes a range of booklets that are used as an aid to:

- classifying growth potential;
- exercising thinning control;
- forecasting timber production;
- estimating the volumes of standing, felled and stacked timber.[1]

Equipped with such items the forester will wish to survey the wood and will make assessments regarding:

- main tree species and the area they occupy;
- crop densities (stems per hectare);
- growth rates (yield class);
- range and average dimensions of trees;
- tree quality;
- workability of the ground (under different weather conditions);
- availability and quality of access routes;

- position of convenient lorry loading points;
- harvesting equipment needed;
- state of local and national timber markets;
- availability of contractors.

It is clear that some level of training is desirable if meaningful assessments of timber production potential are to be made. The average small wood owner will usually need to seek specialist assistance for this work.

Commercial Merits of Broadleaves and Conifers

Depending on site conditions, broadleaved crops in lowland Britain normally grow at the rate of five to ten tonnes per hectare per year and conifers at ten to 25 tonnes or even more. Because of faster growth rates and more regular form, conifers are generally more profitable than broadleaves. Well-managed conifer crops can be worth many thousands of pounds per hectare when the trees are ready to be felled. By comparison, the average standing value of unmanaged broadleaved woodland is unlikely to exceed a few hundred pounds per hectare.

Provided logs are clean, straight, over 18 centimetres in diameter and can be supplied in quantities of at least a lorry load (about 20 tonnes), most species should be easy to market and should fetch good prices.

Occasionally a single tree can be of exceptional value where it meets the needs of some specialized market. But even where good trees exist, many are overlooked because owners do not understand their true market value.

Table 2.1 *Favourable Species for Timber Production*

Conifer	Broadleaved
Larch	Oak (not Turkey oak)
Douglas fir	Beech
Spruces	Sweet chestnut
Pines	Wild cherry
Large yews	Ash
	Walnut
	Sycamore
	Norway maple
	Large fruit trees
	Trees with large burrs

Table 2.2 *Estimation of Whole Trees Required per 20-tonne Lorry Load*

Average Tree Diameter at 1.3 m above Ground (cm)	Approximate No Trees per 20-tonne Lorry Load
5	2000
10	500
15	200
20	80–100
25	40–50
30	28–33
35	18–22
40	13–16
45	10–12
50	8–10
60	6–8

Note: Assumes loads of whole tree lengths

Woodland Restoration

Appropriate management can make neglected woods profitable, in the long term at least, by aiming to improve timber quality.

Depending on the composition and condition of the wood, and the owner's objectives, rehabilitation work might involve:

- improving access;
- improving drainage;
- thinning;
- coppicing;
- singling;
- felling and restocking unproductive areas;
- felling and restocking windblown areas;
- cleaning (removing climbing plants, etc);
- pruning side branches of selected trees;
- planting additional trees where large gaps exist.

IMPROVING ACCESS

The nearer timber lorries can get to the woodland, the better the price for the timber. Field headlands and woodland rides are often too rough or too soft for loaded lorries. Surface grading or stoning to improve access will add considerably to the appeal of a timber sale.

THINNING

Thinning involves the removal of a proportion of the trees in a wood for economic or environmental objectives. It is an important operation that can make or mar a tree crop.

Why thin trees? When growing, trees respond in different ways to the amount of space available. Trees grown too close together produce thin, weak stems with narrow crowns and restricted root systems. Slower growing trees become suppressed and eventually die, and this can result in economic loss. The trees that do survive are worth less, so all prospect of profit may be lost. Thinning aims to improve overall quality by removing poorly formed trees and by providing the chosen ones with sufficient space to maintain good growth rates until the time of the next thinning. It can also provide intermediate returns from the sale of produce.

But which trees should be removed? How many should be removed? When should thinning start? How often should trees be thinned? When should thinning stop? These are all aspects of thinning control – a topic that has greatly exercized the minds (and journals) of foresters since trees were first grown in plantations. Novices should not be discouraged by the seeming complexities of the subject. To a large extent thinning is a matter of observation, common sense and good judgement. In any event, in practice the cumulative volume production of usable timber will be little affected over quite a wide range of thinning intensities. Though a regularly thinned wood will be made up of larger trees, it will not necessarily have a greater volume of timber than an unthinned one. It will, of course, be considerably more valuable.

Some of the most common forms of thinning are considered below.

Selective Thinning

The initial thinning will often be a selective thinning that aims to improve the appearance and the value of the crop by the removal of trees that are:

- dead;
- dying;
- diseased;
- suppressed;
- weak;
- forked;
- defective in some other way.

Open grown trees tend to have short main stems with long, coarse side branches.
Conservation value high; timber value low.

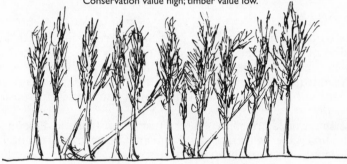

Too close a spacing produces tall, spindly, unstable trees.
Conservation value low; timber value low.

Regular thinning allows selection of the best trees.
Conservation value moderate; timber value high.

Figure 2.1 *Why Thin Trees?*

Subsequent thinnings will generally aim to achieve an even distribution of the best trees to leave a final crop for clearfelling. To achieve this, it may be appropriate to occasionally remove a good tree in favour of one or more poorer ones. In areas with many poor trees, it would obviously be inappropriate to remove them all. Qualities to look for in retained trees are:

- straight stem with absence of forks;
- vigorous growth;
- light and even branching habit;
- healthy crown that occupies roughly one third of the stem.

Line Thinning

Where growth is consistently even and there are no important landscape considerations, line thinning may be employed. This involves the systematic removal of complete rows – typically every third, fourth or fifth row. A first thinning will often combine a selective and a line thinning, the felled rows being used for timber extraction.

Some woods are so dense that inspection is made difficult or even impossible. In these cases it may first be necessary to cut out occasional rows as inspection paths or 'racks', whether or not line thinning is to be carried out.

Crown Thinning

This is a variant form of selective thinning and is particularly useful where the cost of thinning is likely to be greater than the value of the produce. Potential final crop trees are identified and relieved from competition by removing some of the adjacent dominant trees. Harvesting a smaller quantity of larger trees in this way may allow the operation to become profitable.

Thinning of Mixtures

A sound knowledge of the cultural characteristics of tree species is essential for the successful management of mixed-species crops because different species will grow at different rates. Early action may be needed to relieve competition where slower growing species are favoured – though some species will tolerate much more shade than others. This characteristic allows the development of well-structured, mixed-age, continuous-cover woods.

Box 2.1
HOW TO MARK A THINNING

The decision on whether to mark the trees to be removed or the trees to be retained will usually depend on the relative number in each category. If a greater number are to be removed, then marking the retained trees obviously entails less work, and vice versa.

As a general rule, retained trees are marked with spots using special aerosol paints, while trees to be removed are marked with a 'blaze' using a billhook or slasher. Two marks applied to opposite sides of the stem is the usual practice. Until they gain more confidence, beginners may prefer to use a less permanent method that will allow initial mistakes to be rectified. Plumber's tape tied around the trees makes for a safer alternative. It is inexpensive, light to carry, can easily be snapped by hand into required lengths and is clearly visible from all directions.

When to Thin

The economic case for thinning will frequently hinge on whether the operation can be made to pay. Since it improves the quality of the crop, it can be argued that an early thinning at a loss can be profitable in the long run. All too often thinning is left until the timber is profitable, by which time the standing crop may have suffered irreparable damage. It is, nevertheless, disconcerting to be faced with an initial loss situation when considering an operation that is deemed necessary.

It is a fairly straightforward matter to subjectively judge the need for thinning. This occurs when the crown and root development of individual trees becomes restricted. In plantations, this can occur at about 12 to 30 years' growth depending on species, growth rate and spacing. In an overstocked wood the crowns will be much narrower and the stems more slender than is the case for similar trees growing in the open. In extreme cases close grown trees can become so weak that they are incapable of responding to additional space. Ash, oak, pine and larch will rarely respond satisfactorily after being overcrowded.

Subsequent thinnings are normally carried out every four to six years throughout the life of a conifer crop and every five to ten years for broadleaves, but the decision to thin will be heavily dependent on species, age, growth rate, topography and available markets.

Box 2.2
HOW TO MARK TWO THINNINGS IN ONE OPERATION

Given the ability of some modern timber paints to remain visible for many years, it is possible to mark two thinnings in one operation. This can work well in some overstocked crops and neglected woods where management is not meticulous.

First, an even distribution of the best trees is marked for retention with paint spots. These might typically comprise about half of the total number of trees. At the same time, about half of the remainder – those most seriously competing with the chosen ones – are marked for immediate removal with slash marks. In about four or five years' time, when the crop has had a chance to stabilize, the unmarked trees are removed to leave just the paint-spotted original selection.

Thinning Intensity

Trees grown for timber production are generally planted at between 1000 and 2500 trees per hectare. Throughout the life of the crop this number will be reduced in a series of thinnings to a final density of between 60 and 300 trees per hectare, depending on species and rates of growth. As a general rule, about 20 to 30 per cent of the total number of trees (but not the total volume) may be removed in any thinning operation, though again much depends on circumstances (species, condition of crop and time since last thinning). Unless the crop is being thinned to waste (trees are left where they fall), access for extraction vehicles must be accommodated.

Wind Damage

All woods may be damaged by wind as they grow taller but neglected woods are particularly vulnerable. Weak stems can break and trees with poor root systems may blow over in high winds; indeed, storms can cause damage on a catastrophic scale.

The exposed boundaries of woods are generally thinned less intensively than more sheltered areas. Expert advice should be sought before removing leading-edge trees (for example, to widen a ride or to improve boundary design) as this can render crops susceptible to wind damage. Considerable care is needed when thinning an overstocked wood. Reducing the stocking to what is considered a more 'normal' level in a single operation can leave the crop extremely vulnerable to wind damage. Even well-managed woods are more prone to damage in the period immediately following a thinning. The risk can be minimized by thinning lightly and at frequent intervals.

Epicormic Shoots

Epicormic shoots are tiny branches that can develop, apparently in response to thinning, from suppressed buds on the main stem of some trees. They are usually regarded as an undesirable character that reduces the quality and value of timber, particularly oak.[2] They can be removed by pruning.

While undertaking a marking for thinning, the opportunity should be taken to count the number of trees to be removed (a tally counter is a useful aid) and, in appropriate cases, to estimate the volume of timber to be removed.

Table 2.3 *Estimation of Tree Volume from Diameter*

Diameter at Breast Height (cm)	Approximate Tree Volume (cubic metres over bark)	
	Conifers	Broadleaves
5	0.01	0.01
10	0.04	0.04
15	0.10	0.10
20	0.25	0.25
25	0.5	0.4
30	0.7	0.6
35	1.0	0.9
40	1.5	1.2
45	2.0	1.5
50	2.5	1.9
60	3.6	2.6
70	4.6	3.3
80	5.5	4.2
90	7.0	5.0
100	9.0	6.0

Box 2.3
HOW TO ESTIMATE TREE HEIGHT

Apart from direct measurements involving climbing (which is not recommended) or felling trees, height measurements must rely on geometrical or trigonometrical principles. Great accuracy calls for expensive instruments, but a rough estimate of height is possible with some quite basic equipment.

Close one eye and hold a pencil at arm's length so that the point is level with the top of the tree. Adjust the thumb on the pencil so that it is level with the foot of the tree. Stay in the same place and turn the hand in an arc until the pencil is in line with the ground. Note the exact spot on the ground indicated by the tip of the pencil. Walk to this point and measure the distance from there to the base of the tree for a good estimate of its height.

Another method involves the use of a straight rod of about one metre in length. Hold the rod at arm's length and allow the tip to swing towards the face until it touches the cheek. Swing the rod upright again and then walk towards or away from the tree until that part of the rod, from the top of the hand to the tip, is in line with the bottom and the top of the tree respectively. The distance from that place to the tree is roughly equal to the tree's height.

Estimating the Volume of Standing Trees

To gain an accurate estimate of standing tree volume involves quite complex methods of mensuration that are beyond the scope of this book. A rough estimate, however, can be obtained from tree diameter and reference to Table 2.3.

Tree diameter is typically measured 1.3 metres above ground level. This measure is known as diameter at breast height (dbh). A special timber girthing tape, calibrated in centimetres diameter, is carefully placed around the circumference of the tree at breast height and the diameter is directly read off. Girth tapes can be obtained from specialist forestry suppliers and chainsaw retailers. In the absence of a girthing tape, homemade callipers can serve almost as well, though it should be remembered that not all trees are cylindrical so the average of two measurements at right angles is the usual practice.

In referring to Table 2.3 it should be noted that differences in tree shape may cause variations of 50 per cent or more. Volume of branch wood is not included.

Figure 2.2 *Estimating the Height of a Tree*

Estimating the Volume of Felled Trees

The contents, in cubic metres, of felled timber can be found by reference to *Forestry Commission Booklet 39*.

For the layman, it is difficult to imagine tree dimensions for any given volume. Figure 2.3 illustrates, for a number of logs of approximately one cubic metre in volume, a range of equivalent mid-diameters and lengths.

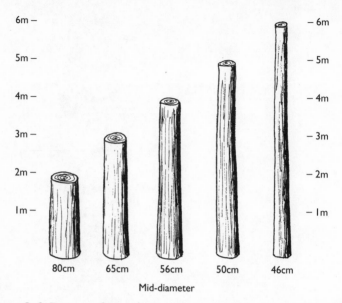

Mid-diameter

Figure 2.3 *Range of Log Sizes Approximately One Cubic Metre in Volume*

FELLING

All tree crops grown for timber production will reach a stage when they have to be felled. For conifers, this will not usually be until the trees have attained an average of at least 30 centimetres dbh and preferably larger. Typical felling ages (rotations) for conifers are in the range of 45 to 65 years. Broadleaves will normally not be felled until average diameters reach 45 centimetres or more. The rotational age for faster growing broadleaves, such as ash, sycamore, wild cherry and sweet chestnut, is generally in the range of 60 to 80 years, while that for some of the slower growing species, such as oak and beech, can be as long as 120 to 180 years. Provided crops have been properly maintained and regularly thinned the final felling should bring in a substantial income.

Organization and control

The dangers associated with woodland felling operations are considerable. Accidents, even deaths, occur with sickening regularity in the forestry industry. Work should never be undertaken by untrained, ill-equipped personnel. Tree felling by chainsaw requires

Box 2.4
HOW TO MARK TREES FOR FELLING

Trees to be felled should be identified in a distinctive and semipermanent way, bearing in mind a year or more may elapse before the work is completed. The trees need not be individually marked provided the boundaries of the area are clearly identifiable and can be accurately described on paper. Even so, it may be prudent to mark the boundary of the clearfell area on the ground. More usually (and particularly for small felling areas), trees are marked with slash marks as in a thinning.

Where high value trees are involved they should be individually numbered with timber marking paint to correspond with an inventory of volume estimates.

the highest level of skill and attention to safety. Trees on wet, steep slopes can pose particular problems for non-professionals. The efficiency of subsequent operations can be adversely affected by amateur felling. Wherever doubt exists, skilled contractor fellers should be used. The public should be warned of the hazards at all points of entry to the wood, and if anyone strays into the working area, work should stop until they leave. It may be necessary to divert visitors around the working area.

Good organization and control are necessary to ensure an efficient and orderly operation and to ensure the site is left in an

Box 2.5
HOW TO FELL A SMALL TREE

Clear any growth from around the tree and plan an escape route should a rapid exit become necessary. Decide where you want the tree to fall and check to see if there is any likelihood of the tree becoming lodged or of large branches falling. Realize that if the tree is leaning in an unfavourable direction there will be less or no chance of it falling where required.

Consider the length of the tree and ensure that there are no impediments, such as large stumps, boulders or fallen trees, over which the severed tree may fall and suddenly spring back (or break its back). Watch out for power lines and inform the appropriate authority if there is any danger of the tree falling near the line.

Cut, using axe or saw, a V-shaped 'sink' at the base of the tree on the side where the tree should fall. Sawing is then started on the opposite side of the tree at a slightly higher level and is continued until the tree starts to fall.

1. Use saw or axe to cut 'V' shaped sink on side tree is to fall

2. With saw level, cut from opposite side at slightly higher level

Figure 2.4 *Method of Felling a Small Tree*

acceptable condition ready for replanting. Extraction routes, conversion points and stacking or loading bays should be identified before felling commences. It is essential that the trees themselves are felled carefully. Inexpert felling, particularly of mature broadleaves, can cause considerable damage to the timber with subsequent loss of value. Compared with normal felling situations, the harvesting of windblown trees is inherently more difficult and slower in various ways.

Season of Felling

Felling is generally possible throughout the year, though the bird nesting season, from March to July, should be avoided whenever possible. Conifers may be felled at any time, although staining may occur to logs left lying for more than six weeks or so during the summer. Pine species should not be left lying as they are likely to harbour insects harmful to conifers growing nearby and they suffer from timber staining. High-quality broadleaved trees should be felled in autumn and winter when the sap is down. It is essential for logs of certain species (particularly ash, beech and sycamore) to be extracted from the site as soon as possible after felling, otherwise quality may suffer. Sporting considerations will often affect the timing of felling operations.

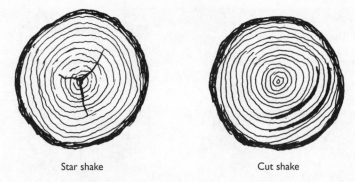

Star shake   Cut shake

Figure 2.5 *Shake Splits in Timber*

Shake

'Shake' describes the splits that can develop in the wood of trees. It is the cause of serious timber devaluation in several species, but particularly in oak. Oak logs, otherwise of veneer quality, can be reduced to firewood quality by the presence of severe shake. The aspects of tree growth that are believed to be associated with the development of shake are complex and not fully understood.[3] Certain regions of Britain have reputations for producing a high incidence of shake. Environmental factors are thought to influence the condition. Gravels, stony soils, waterlogged soils and very windy sites should be avoided when planting oak. Moist soils, deep soils and those with a high clay content are considered more desirable.

EXTRACTION

As with felling, the extraction of timber can involve operators in considerable risk of injury. Health and safety aspects will loom large. Again, the work calls for training. In planning extraction operations:

- Mark out routes to minimise ground compaction;
- Avoid needlessly criss-crossing the site with heavy machinery;
- Avoid extraction when the ground is wet;
- Consider the use of horse extraction on sensitive sites;
- Consider onsite conversion using a portable sawmill where extraction is over long distances.

A wide range of extraction vehicles, including some agricultural tractors, have been successfully used in small woodlands.[4] Some of the more common extraction methods are discussed below.

*Timber from thinnings and fellings can be utilized for
many purposes*

Tractor Skidding

Skidding involves the dragging of timber along the ground to the
lorry loading point. In most circumstances the timber needs to be
partly supported if it is not to be contaminated by dirt and stones
along the whole length of the log. This can lead to sawmilling
problems and may affect prices. It can also result in unacceptable
levels of ground damage. The reinstatement of rides and tracks
after a poorly executed skidding operation can take much of the
profit from a harvesting operation.

Tractor Forwarding

Forwarding is the process of moving cut-to-length timber to the
lorry loading point on trailers. A wide range of timber trailers,
powered and non-powered, is available. Tractors and trailers must
be properly matched and suited to the work programme for
optimum performance. For extraction distances of about 250
metres or more, forwarding is usually more economical than
skidding.[5]

Horse Extraction

Until about the middle of the 20th century the horse was one of
the main methods of timber extraction. However, the introduction

of tractors with the potential for higher outputs and reduced costs caused a decline in the use of the horse. Today, with the renewed interest in small-scale woodlands, the use of the horse, in conjunction with various devices designed to keep timber clear of the ground, is growing. Horses can be used to extract large logs in environmentally sensitive areas and where conditions limit other methods of extraction.[6] Their use has been shown to be cost-effective in appropriate circumstances.[7]

The Iron Horse

The iron horse is a small pedestrian-controlled extraction machine driven by two cleated rubber tracks. Engine power can be either five horsepower (hp) or seven hp and there is an optional trailer and winch. Similar methods and outputs apply to timber extraction by the iron horse and the traditional working horse.[8] The machine is compact, robust, easy to operate and manoeuvrable. It has a role in small-scale and non-commercial extraction, but training in its use is essential.

Other Extraction Systems

Timber extraction in mountainous country may be beyond the scope of tractors and horses. The work must then rely on equipment such as cable cranes, skylines, flumes or helicopters. Only in exceptional circumstances, or where substantial quantities of timber exist, can such high-capital cost systems be justified.

CONVERSION

Conversion involves the cutting of felled trees into product sizes. The operation calls for a range of specialized equipment, which might include:

- chainsaw;
- retractable measuring tape;
- timber tongs;
- log-turning jack;
- circular saw bench;
- wood chipper machine;
- bark peeling machine;
- wood splitter device.

Before the trees are felled, it is possible (using the assortment tables in *Forestry Commission Booklet 34*) to judge the relative quantities of the most lucrative products that can be cut from the trees. Table 2.4 indicates a range of produce associated with softwoods and hardwoods.

Table 2.4 *Assortments of Conifers and Broadleaves*

Conifer Assortment	
Sawlogs	The value of conifers is mostly dependent on the amount of sawlogs they will yield. Parcels with an average size over 30 centimetres dbh will contain a significant amount of sawlog material and are generally in high demand.
Fencing Bars	There is often a call for intermediate sizes for 'bars' for pallet and lap-fencing manufacture.
Stakes	Smaller sizes will go as stake 'blanks' for processing into round, pointed fencing stakes.
Pulpwood	The topmost end of the tree is useful for conversion into pulpwood lengths.
Broadleaved Assortment	
Veneer Logs	Large, clean, straight, cylindrical hardwoods can fetch seriously high prices for veneer logs but quality has to be exceptional. Britain has no veneer mills. The principle importing country is Germany.
First-Grade Logs	Trees over 30 centimetres dbh should be readily saleable provided the lower parts of the stems are straight and free of major branches. First-grade logs, primarily for the furniture trade, are always in high demand.
Second-Grade Logs	These are logs with minor defects. They will usually find a market.
Pulpwood	Poorer trees and those less than about 18 centimetres dbh are commonly converted into pulpwood.
Firewood	In favourable circumstances firewood can be a more lucrative outlet than pulpwood.

*The use of a portable sawbench can add value to
woodland produce*

Mobile Bandsaws

Mobile bandsaws allow for the conversion of timber where it is
felled – in the wood. From each log different widths and thicknesses can be precision-cut, minimizing wastage and optimizing
profitability. By hiring the services of a specialist contractor with a
modern mobile sawmill, the woodland owner is given the option of
producing a range of quality sawn products, for home use or sale.
Some bandsaws weigh no more than one tonne and can be towed
into most woods by a four-wheel drive vehicle, with setting-up
time from arrival to first cut as little as ten minutes.

The thin blade ensures efficient conversion, so that even when
converting as few as two or three logs, use of a mobile bandsaw can
turn a potential operating loss into a profitable undertaking.
Furthermore, since large-scale extraction equipment is not required,
the converted material can be carried easily from the woods by light
vehicles, reducing the potential for environmental damage.

Bandsaws can convert around eight to ten cubic metres in a
day depending on planking dimensions and log size. Logs as large
as 90 centimetres in diameter by 6.5 metres in length can be cut
into thicknesses down to about 2 millimetres, thin enough for
veneers. Square-sided beams for building renovation, large planks

of furniture quality, and 'waney-edged' boards for rustic furniture and cladding can all be produced from appropriate logs. The cost of milling is dependent on:

- log size;
- log cleanliness;
- the dimensions required;
- site organization.

Where the timber is to be sold, profits of 200 to 300 per cent are not uncommon, especially when sawing large-sized timber or quality hardwoods. The sawmill operator can often help with marketing the sawn material. The basis of the contract should be clearly understood before work commences. Payment may be based upon a day rate, a price per log, a price per cubic metre, or on a rate for yield of sawn material.

Planning the Work

All operations should be designed to avoid unnecessary double handling. Before any felling takes place the sawyer should visit to discuss requirements. It is the woodland owner's responsibility to decide upon the dimensions of timber to be sawn. If access is reasonable, logs can be processed 'at stump'. However, where several different products are to be sorted into separate stacks, it can be more efficient to extract whole tree lengths to a woodland depot for conversion. Wherever work takes place, trees are best crosscut into required lengths or multiples of required lengths before the sawmill arrives. They should be as straight as possible and any crooked sections should be cut out and discarded. Any time the sawyer has to spend crosscutting will have to be paid for. Remember, short logs produce less timber but require just as much handling as longer lengths.

Where conversion is to be at a depot, the logs should be lifted and transported to the site rather than skidded on the ground. Small stones and dirt lodged in the bark will have to be removed before sawmilling, or else the logs will have to be peeled. If trees have been used to support fences they should be rejected or marked clearly to show the side that may contain staples or wire. Blades are very expensive and most sawmillers charge for damage caused by metal inside timber. A metal detector is a useful tool in these circumstances.

For most species, it is best to keep the time between harvesting and milling to a minimum to reduce the incidence of timber degradation during log storage; however, oak can be safely stored in the log before conversion to reduce any tendency to shake when sawn.

The chosen work site should be reasonably level and spacious. If possible, there should be sufficient room for the sawmill to set up on either side of the stack. The operator can then move position if the wind blows the sawdust in an awkward direction. Logs should be laid, butt-ends together, on bearers to reduce timber degradation and to ease handling. They should not be stacked more than two logs high for safe and efficient work. The aim should be to handle the timber only once, off the saw and into the drying location. It is worth remembering that sawn material will take up more space than logs.

Children, farm animals and pets must be kept clear of the worksite. Adequate insurance cover is advised, particularly if the owner's own labour or support equipment is being provided.

SEASONING TIMBER

It is essential that sawn wood be properly seasoned. Poor practice can ruin the timber. A lean-to on the north side of a barn makes a perfect shelter. The site should be:

- easily accessible;
- level;
- well ventilated;
- protected from direct sunlight;
- protected from the prevailing wind;
- covered.

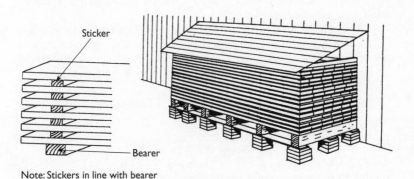

Note: Stickers in line with bearer

Figure 2.6 *Air-Seasoning Timber*

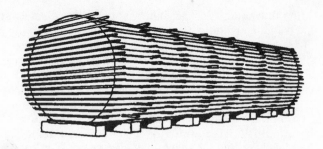

Figure 2.7 *A Log Sawn Through and Through, Set on Bearers, with Stickers Separating each Tier*

The bottom planks should be off the ground on bearers. Wooden stickers separating each tier of sawn timber are vital. They should be about 15 to 20 millimetres in thickness and kept about 0.4 to 0.5 metres apart. They must be positioned perfectly in line with one another and the bearers. Oak should never be used since it can stain the wood and is so hard that it may leave indentations on the planks. Stickers made from conifer wood such as larch are ideal.

The maximum width of stack should be about 1.8 metres. Within reason, the higher the stack the better; the increased weight will help resist twisting and bowing so the best-quality timber should be kept at the bottom. With small quantities, banding or weighting can be used to reduce any warping effect. To retard the rate of drying and to minimize the risk of planks splitting, the ends can be sealed using an arboricultural sealant or household paint.

Traditional air-drying of wood can reduce moisture content to about 20 per cent, but kiln-drying can reduce it to about 10 per cent. Kiln-dried hardwoods can be worth up to three times as much as unseasoned timber.[9]

Until relatively recently, kiln-drying has been fairly inaccessible to small-scale users since much of the available equipment was either expensive or complex.[10] However, there are now a number of simple, robust kilns on the market. They can be supplied ready to run, require no building or permanent foundation and, in most cases, a single-phase electric supply is all that is needed.

SELLING TREES

Having first given thought to the possibilities of utilizing the timber themselves, owners must decide whether to sell their trees standing, felled or converted. Inexperienced sellers of trees are generally

advised to sell their material standing the purchaser is then responsible for felling, extracting and converting the trees. This method involves the owner in the least outlay, work and commercial risk, and it leaves the detailed marketing to the purchaser, who will know the requirements of the main sawmillers and processors, which products to cut from the trees, and who will identify where various species and grades can be most profitably directed.[11] Furthermore, the purchaser accepts the risk of any defects in the timber or breakages during felling.

Valuing Trees

A purchaser's offer for standing trees will be substantially less, volume for volume, than the value of the converted product. Valuation will be based upon:

- species on offer;
- total volume;
- individual tree volume;
- tree quality;
- ease of access;
- estimated felling costs;

- estimated conversion costs;
- distance from wood to main market;
- current state of supply and demand in the timber trade.

The distance the timber has to be moved after felling, from stump to lorry loading point, has a significant effect on value and saleability. A distance exceeding 500 metres is likely to lead to a marked reduction in the price offered.

Not infrequently, where small lots of low-quality trees are concerned, harvesting costs will exceed the value of the end product, resulting in a net cost to the owner. It is generally not worth spending a great deal of effort attempting to sell parcels where the average diameter of the trees is less than, say, 12 centimetres dbh. It is probably better to wait a few years while tree dimensions increase. (This does not mean that first thinnings can be delayed indefinitely – crops can be ruined by delaying the first thinning too long.) As a general rule:

- Small parcels will fetch a reduced price because setting-up costs will be proportionally more expensive.
- Very small parcels of less than one lorry load can be extremely difficult to sell unless comprised of particularly valuable material.
- A mixed-species parcel will attract a lower price than a parcel containing a single species.
- Small-sized broadleaved trees are generally lower in value than their conifer equivalent.
- Conversely, large broadleaves producing high-grade timber will usually achieve substantially higher prices than will the largest sizes of conifers.
- First thinnings of both conifers and broadleaves are frequently uneconomic.
- Subsequent thinnings should prove progressively more profitable.
- A clearfell parcel will be more saleable than an equivalent volume from thinnings.

Methods of Sale

Timber sales are made by auction, tender or negotiation. Which method is chosen depends largely on the value of the parcel.

Timber Auction

An auction is appropriate mainly where substantial volumes or higher-value parcels are involved. The recent introduction of electronic auctions into the world of timber sales may provide a suitable opening for the sale of smaller lots by auction. Vendors do not need to be connected to the system.

Competitive Tender

Timber lots can be sold by competitive tender, either public or restricted to a few local merchants. A disadvantage of the tender system is that some considerable time can elapse between set-up and the receipt of offers.

Negotiation

For small-scale and lower-value parcels the benefit of a competitive sale may not be large and negotiation with a selected timber merchant or contractor is usually much more convenient. At least three offers should be sought.

The seller should take a realistic view of the interval between the planning of a sale and the actual start of felling. Time is needed on the seller's side to mark the trees to be felled, to prepare a felling licence application or a grant scheme application and to negotiate the sale. Time is also needed by the Forestry Authority to process the application and by the purchaser to move his equipment to the site. The overall time can be in the order of six months.

Woodlots

In 1995, the Forestry Authority launched a free advertising magazine called *Woodlots*, which is distributed bi-monthly to thousands of timber merchants, sawmills, furniture-makers and other wood users. Wood producers can place small adverts, free of charge, in the hope of selling parcels of wood or woodland products and getting a market for them – anything from one tree to a large quantity of logs. Adverts are also placed by those actively looking for timber. For further information contact the nearest Forestry Authority office.

Basis of Sale

Standing trees are commonly sold in the following ways:

- agreed price – payment on a lump-sum basis or as instalments on specified dates or as and when defined quantities are completed;
- by weight – payment per tonne collected (from weight tickets issued from a public weighbridge);

- by stacked volume – from a measure of produce at the roadside;
- by tree volume – from a measure after felling;
- per tree;
- by number of pieces of each product produced.

Sale Particulars

Whatever the method of sale, some kind of crop description must be prepared. The particulars can range from a simple to a detailed description. In many cases the following will suffice:

- statement of species;
- statement of whether clearfelling or thinning;
- estimate of average dbh;
- number of trees;
- estimate of volume;
- statement of how trees are marked;
- a map showing area to be worked and authorised access routes.

Written Agreement

An agreement to sell standing trees should be recorded in writing. A simple contract should be prepared for signature by both parties, to the effect that the trees described in the sale particulars (and delineated on an attached map) will be purchased, felled and removed by the purchaser. Though it may form part of the sale particulars, it is not recommended that a volume estimate be included in the written agreement because of the imprecise nature of many methods of volume estimation. Standard contract forms may be available from local Forestry Authority offices. The agreement will normally comprise:

- the basis of sale (as above);
- the terms of payment (as above);
- starting and completion dates;
- schedule of condition of fences, walls, roads, rides, etc;
- extent to which the purchaser will be responsible for damage to fences, walls, roads, rides, etc;
- a requirement that the purchaser holds sufficient insurance to be able to meet any damages;
- methods of disposal of lop and top.

The seller can choose to retain the right to stop the use of any machine or method of working that is causing, or is likely to cause, damage to property, but it is important not to impose too many conditions on the sale. Where constraints lead to inconvenience and expense to the purchaser, then inevitably a reduced price will be obtained and ultimately the saleability of the parcel will be affected.

COOPERATIVE WORKING

There is a need to distinguish between cooperatives and cooperation. The possibility of assembling woodland owners into some form of cooperative to harvest and market produce collectively has been suggested as a solution to the problems of small woodland working.[12] Formal woodland cooperatives have been tried but have rarely proved successful. Such initiatives do not seem to meet the needs (or the temperaments) of small woodland owners. On the other hand, cooperation on a local scale between two or three owners is well worth while. They might agree to work together to:

- exchange information and share experiences;
- combine all woods under one management plan;
- share consultancy and supervision costs;
- share machinery hire costs;
- lend each other equipment;
- share skilled labour;
- employ the same contractor;
- combine sales of small timber lots.

COPPICING

Coppicing is a traditional woodland management system that exploits the ability of most British broadleaved trees to produce new growth from the cut stump or 'stool'. For thousands of years coppicing was the most common form of woodland management over much of Britain. Substantial areas of woodland were worked to provide regular supplies of a wide range of products, but as markets declined many copses were abandoned or converted to plantation forestry.

Today, changing lifestyles are producing new marketing opportunities and greater environmental concerns are altering the values that people place on both the countryside and their purchases. At the same time, there has been a revival of interest in woodland crafts, with many people learning old skills and looking for ways of putting them to use. Although few of the traditional large-scale markets for coppice products remain, demand for some items (for example, wattle hurdles for garden screens) currently outstrips supply. Production of the following products is worth considering:

- firewood;
- kindling wood;
- charcoal;
- thatching spars;
- hurdles;
- rustic poles;
- bird tables;
- garden furniture;
- rose arches;
- pergolas;
- screens;
- fence stakes;
- walking sticks;
- pea sticks;
- bean sticks;
- hedging materials;
- race course jumps;
- besom brooms.

Well-managed copses create ideal cover for game, rejuvenate ancient landscapes and provide many hours of healthy and enjoyable activity for coppice cutters, pole-lathe turners, greenwood furniture-makers, hurdle makers, charcoal burners and others.

Environmental Benefits

Much of the diversity and richness of the country's woodland wildlife has thrived under the coppice system for thousands of years. Periodic cutting boosts the growth of a wide variety of plants, while trees in their various stages of growth provide, in a relatively small area, a great variety of habitats for animals, birds and butterflies. On the other hand, managed coppice does not have any dead trees nor does it have the more shaded, damp conditions of neglected woods that can provide a niche for a range of rare and delicate species unable to survive the impact of regular site clearance. It is important, therefore, that some parts of a wood, particularly those that may contain standing and fallen deadwood, be left unworked to serve as reservoirs of biodiversity. It is useful, too, if some trees can be retained to grow old and die naturally.

Coppice can be used for many useful articles

From a conservation standpoint, coppicing is most advantageous in:

- ancient and seminatural woods with a history of coppicing;
- mixed native species rather than single-species stands;
- intensively managed arable districts with few woods;
- woods that have been cut during or since the Second World War.[13]

Coppice Systems

Pure Coppice

This system involves a single-species crop. The most commercially important species, certainly in central and southern England, was hazel, usually worked on a six- to 12-year cycle. Left longer than this, hazel has few uses other than as firewood. Sweet chestnut, oak, ash, lime, field maple, birch and beech were also coppiced, either pure or in mixtures with other species. Even when retained well beyond the normal time of cutting, most of these other species can still be marketable.

Many small woods are overstocked due to years of neglect

Mixed Coppice

Many of our existing mixed coppice woods are the result of regrowth following clearfellings during the two World Wars. Subsequent management has generally been lacking so that many of these woods are now heavily overstocked with trees. Some are virtually impenetrable.

These neglected woods can be unstable in high winds, are likely to produce little or no profit and are of only limited value for wildlife. Nevertheless, appropriate management can improve them greatly.

Coppice-with-Standards

This system comprises an even-aged coppiced underwood and an uneven-aged overstorey of 'standard', or large-size trees grown for timber. The separate components seldom achieve full potential. The coppice tends to be less productive due to competition for light, moisture and nutrients, while the standards have so much room they tend to develop shorter boles and rougher branches than close-grown individuals. One of the most common coppice-with-standards systems involved the use of hazel coppice under oak (or mixed oak and ash) standards.

The standards should be fairly evenly spaced with their crowns occupying about 30 to 50 per cent of the ground area. Sufficient light must reach the coppice to enable it to grow vigorously. This could mean up to 100 standards per hectare, depending on size, but the most productive sites might support 180 standards or more per hectare.[14] In a 20-year coppice rotation wood, the number of standards per hectare, immediately after a cutting, might be:

- 100 trees at 20 years old;
- 50 trees at 40 years old;
- 25 trees at 60 years old;
- 10 trees at 80 years old.

At each coppice cutting most of the older standards and a number of standards from each of the younger age classes (typically the poorest ones) are felled and utilized. At the same time a number of new trees must be planted, or some naturally regenerated trees recruited, to provide a selection of trees from which future standards will ultimately be chosen.

Short Rotation Coppice

Otherwise known as SRC, arable coppice or biomass, this is a more recent form of management that involves the growing of willow or poplar on surplus farmland on very short rotations. It is more an agricultural, rather than a forestry, enterprise and is outside the scope of this book.

Box 2.6
HOW TO CUT COPPICE

Depending on the size of the stems to be cut, a billhook, bowsaw, axe or chainsaw is used. When using a chainsaw, to avoid snags, all small whippy growth from around the base of the stool should first be cut away with a billhook.

With most species the cut should be made at the previous level without greatly increasing or decreasing the height of the stool. With hazel, the cut is generally made as low as possible irrespective of the previous level.

Stems are removed one at a time, working in a spiral fashion, from the outside in, towards the centre of the stool. A sloping cut sheds rainwater – though this may not be a crucial factor in reducing incidence of decay. Stools are then cleaned up, cutting off any split wood and brushing sawdust off the cut surfaces.

Cutting Coppice

Traditionally, coppice is cut between October and March when:

- birds are not nesting;
- there is no foliage on the trees;
- working conditions are likely to be more amenable;
- there is a full season's growth ahead for the new shoots.

In practice, coppice can be cut successfully at any time of the year, though late summer cutting may not allow sufficient time for new shoots to harden-off before the onset of early frosts. Furthermore, summer-cut wood is less durable than winter-cut material.

Compared with the growth of a planted tree, the initial regrowth of a coppice shoot is very vigorous. Oak will commonly reach one metre and ash, sycamore and sweet chestnut may grow as much as 2.5 metres in the first year. Even greater height increment may occur in the second year, but thereafter growth resembles that of a planted tree of the same species.

Restoring Derelict Coppice

Many thousands of hectares of derelict coppice now exist in Britain. Provided the trees are no more than 60 to 70 years old, resumption of coppicing is usually successful. Regular coppicing can then be carried on indefinitely and the yield will be maintained without the use of fertilizers.

The restoration of derelict copses can be an expensive exercise, unless the material can be utilized (for example, for firewood or

---

Box 2.7
HOW TO GROW NEW TREES BY LAYERING

When cutting coppice, occasional stems are left uncut ready for layering. The operation is carried out in late winter. It involves bending over a stem, of about three to five centimetres diameter at the base, and pegging it firmly to the ground in the gap. The stem is lightly covered with soil. Roots develop below the shoot and this 'new' plant is recruited to fill the gap. Several species will respond to layering, including hazel, lime and sweet chestnut. With hazel, some workers will twist the bark to bruise it at the point to be pegged.

Layering can be a hazard to walkers, particularly on shooting grounds.

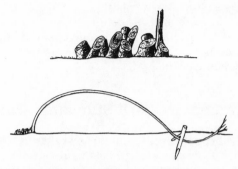

Figure 2.8 *Method of Layering*

charcoal production), in which case it may be possible to have the work done for nothing. Even when properly managed, coppice is unlikely to make owners rich, but it can offer a modest income for a small outlay. Factors that can influence restoration include:

- age;
- condition;
- stocking levels;
- presence of rabbits and deer;
- available expertise;
- availability of markets;
- availability of grants;
- value as game cover;
- value for other wildlife.

Unwanted brash can be left where it falls, but where sporting interests prevail it is best gathered into 'windrows' for easier access by beaters and dogs.

The spacing of stools need not be as close as initial spacing in conventional plantations because each stool produces many shoots. The longer the rotation, the fewer the number of stools needed per hectare. A well-stocked mixed broadleaved copse should have a stocking of about 1000 stools per hectare. Where stocking is less than about 800 stools per hectare it is usual to restock the larger gaps, after the coppice has been cut, by planting, recruiting self-sown trees or layering.

Where planting is necessary (for example, to make changes to the composition of the wood), locally native trees should be used whenever possible. Remove any competing vegetation, preferably by herbicide spot-application. Healthy plants, well furnished with roots and 0.5 to one metre in height, should be planted at 2.5 to three metres spacing. Newly planted trees are generally not coppiced until they are at least seven years old, by which time they should have developed a reasonable basal stump and root system. Competing weeds may need to be controlled to ensure continued vigorous growth.

Restoring a Coppice-with-Standards Wood

Restoring an old coppice-with-standards system is not a simple matter. As years go by, the system falls more out of step. No new standards are established, mid-term trees go unthinned, the understorey becomes weakened by increasing shade, and timber quality declines as the economic age for felling is passed. Restoration calls for a thorough understanding of the system. Many more standards than usual need to be removed at the first coppice cut, but the sudden removal of too many trees can result in wind damage. No standard prescription can be given – each wood must be individually considered and the advice of someone with coppicing experience will be essential.

Protecting Coppice

The control of browsing and grazing animals (particularly rabbits and deer) is as essential in coppice crops as it is in any new planting, though the rapid growth of coppice limits damage very largely to the first couple of growing seasons.

Training

At a time when most industries are highly capitalized, the coppice trades are one of the few areas where the novice can start up without the need for expensive equipment. However, training is essential: while it is possible to cut managed 'in-cycle' coppice using only hand tools, restoring long-neglected copses, felling standard trees and any use of chainsaws is much more hazardous work calling for specialist techniques.

POLLARDING

Pollarding or 'polling' involves the regular cutting of a tree's branches well above the ground – typically at a height of two to three metres. Ancient pollards are an integral part of the historical landscape of Britain but, as with coppicing, the practice has been neglected and there are now many old pollards with very large branches, and some with hollow trunks or 'bollings' that have not been cut for 50 to 150 years.[15]

Almost all species of broadleaved tree (and even some conifers) will form pollards.[16] The most common species encountered are:

■ oak;
■ ash;
■ beech;
■ hornbeam;
■ willow.

Other species less commonly found include:

■ field maple;
■ elm;
■ black poplar;
■ crab apple;
■ sycamore;
■ whitebeam;
■ small-leaved lime;
■ large-leaved lime;
■ hawthorn;
■ holly.

Historically, pollards were created to produce repeated crops of small-sized wood above the reach of browsing animals. The produce was probably put to much the same use as that of coppice, though young, tender shoots and foliage were also cut for animal fodder. Today, pollarding is mainly carried out for landscape and conservation reasons.

Pollarding was a treatment typical of parks and commons, and the existence of old pollards may be evidence that an area was once managed as wood pasture. Pollards are also found on the banks of streams, where their main function is to prevent bank erosion, and on woodland edges, where their longevity led to their use as markers, particularly on parish boundaries. Trees were traditionally repollarded every five to 35 years depending on circumstances, but many trees have not been repollarded for such a considerable time that the bollings are no longer able to support the weight of their overgrown crowns.

All branches
cut close to
previous level.

(a) Pollarding technique for trees regularly pollarded every five to 35 years

Short stubs left.
Young twiggy growth
retained.

(b) Pollarding technique for old neglected pollards with some young growth occurring low down on branches

Longer stubs
left.
One branch
retained.

Retained branch
cut some years later.

(a) Pollarding technique for old neglected pollards with no young growth low down on branches

 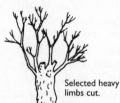

Selected heavy
limbs cut.

Additional
branches
cut after
3–5 years.

Last of the
old branches
removed
a few years
later.

(a) Alternative pollarding technique for old neglected pollards with no young growth low down on branches

Figure 2.9 *Repollarding Techniques*

Box 2.8
HOW TO RE-POLLARD TREES

Repollarding is effectively the felling of trees while the tree surgeon is standing two to three metres up in the air. It is dangerous work and there is no
substitute for proper training. A contracting pollarding specialist is usually
the best option.

The recommended practice for old pollards is to cut the branches a
little way above the branch collar, leaving a stub. The larger the branch
diameter, the longer the stub. Cut at a similar angle to the previous cut.
Retain as much live material as possible. Where little or no young growth
occurs, leave one branch uncut for a year or two to carry on photosynthesis while the new shoots develop. Where possible, choose a branch near the
centre to help keep the tree stable.

A less traumatic treatment is to selectively thin the crown, removing
large structurally unsound branches and retaining smaller ones. Keep the
crown well balanced throughout the exercise. Future cuts should progressively remove the remaining branches. Taking the process slowly over several
years is probably the ideal solution and is particularly recommended for
beech. Proprietary wound treatment paints should not be necessary. Where
a number of old pollards exist, it is sensible to cut only a proportion in any
one season.

Repollarding

Maintaining old pollards helps to preserve traditional landscapes
and contributes to wildlife conservation. The cut surfaces will often
contain irregularities where rainwater collects and where pockets
of deadwood may form. Various fungi may be present and habitats
may exist for many invertebrates and hole-nesting birds. Fungal
infection and the activities of wood-feeding insects are circumstances the pollard can live with for centuries and are not indicative
of imminent death. Indeed, rotting heartwood can be beneficial in
providing resources for the tree's roots to feed on. The bark provides
a long-lived substrate for the establishment of lichen communities,
mosses, ferns, and other epiphytes, and old riverside pollards can
provide nesting sites for ducks and even otter holts.

Trees recently pollarded are much less likely to be damaged in
high winds than old neglected ones. The practice can also be used
to prolong a tree's life – regularly pollarded trees can survive for
centuries.

Box 2.9
HOW TO CREATE NEW POLLARDS

New pollards are created for continuity. Maiden stems are usually chosen but singled coppice stems can also be used. New pollards are best made from young trees with a main stem at about ten to 15 centimetres dbh.[20] Older and larger trees may not survive the truncation process.

Newly pollarded trees need abundant light. Overhead shade and competing side growth must be cleared if regrowth is to flourish. The cut should be at least 30 centimetres above the reach of browsing animals. Where possible, leave branch stubs protruding to encourage new shoots to grow. Retaining any lower branches for a year or two will increase the chances of survival.

Creating new pollards means new, long-term maintenance liabilities. Owners should be certain that they want pollards before creating them. Note that elm generally responds well to being newly pollarded. The small branches produced may be less susceptible to colonization by elm bark beetles.

Prospects of Success

There are obvious differences between the regular repollarding of a tree every few years and the repollarding of one that has remained uncut for a century or more. No attempt should be made to repollard ancient trees without specialist advice. Success can be inconsistent and unpredictable; the more decrepit the tree, the greater the chance of failure.[17]

Species differ in their tolerance to repollarding. The effect on oak, for instance, has been variable, and beech often dies. The repollarding of hornbeam, ash, black poplar and willow is said to be usually successful. Work on ancient beech pollards has shown that regrowth occurs most readily when part of the crown is retained,[18] and the survival rate is far greater when stubs exceed 0.3 metres in length than with shorter stubs.[19]

Trees that receive plenty of light appear to have a better chance of survival, and trees with twiggy burrs and those with plenty of young growth low down on the branches are more likely to respond well. Pollarding is best avoided between the time of bud burst and mid-summer when the tree's reserves may be depleted due to the flush of new growth. Cutting should not take place in the bird nesting season. Better success rates may be achieved by lopping in January or February than during autumn and early winter. Pollarding should be avoided in a drought year or in the year following a drought.

Successful regrowth consists of many shoots arising in a vigorous bushy mass at the top of the bolling. These branches self-thin so that as the branches grow larger there are correspondingly fewer of them.

Keeping Records

There is not a lot of written information about pollarding: it would, therefore, be helpful to researchers if records were kept of the location, species, diameter, height, age and date of cutting.

Grants

Grants for pollarding may be available from local authorities or from MAFF.

FIREWOOD

Wood can be burnt in the form of logs on open fires, or as logs or chips in a wide range of modern stoves, burners and boilers to provide radiant, convected, ducted-air or hot water central heating. Wood is a renewable source of energy that produces virtually no smoke and no acid rain when burnt in an efficient wood-burning stove or furnace. It has a low ash and sulphur content and it represents unburnt fossil fuel. There is a considerable current market for logs that could provide a key to the management of many of our small broadleaved woods.

Home Firewood Production

Restoring small woods for the purpose of home firewood production should appeal to environmentally conscious owners since it provides:

- self-sufficiency;
- reliability of supply;
- personal control;
- freedom from price fluctuations;
- no VAT charges;
- additional sporting, amenity, conservation and shelter benefits;
- an excellent way to keep fit.

Heating requirements can vary greatly with:

- the size of building;
- the degree of insulation;
- the type and efficiency of the heating system;
- the needs of occupants.

Heating a typical three bedroom house for a year might use around:

- 36,000 units of electricity;
- 1200 therms of gas;
- 600 gallons of oil;
- 4.5 tonnes of coal;
- eight to nine tonnes of air dried wood.

Working a Firewood Copse

Although large trees, branches and tops can be cut up and used as firewood, more manageable sizes can be produced on a sustainable basis by coppicing.

A well-managed wood containing a full stocking of mixed broadleaved trees should produce about three tonnes of air-dried wood per hectare each year (about two tonnes of green timber is

---

Box 2.10
HOW TO CUT CORDWOOD

Cordwood is the material derived from the branches of broadleaved trees. For the purposes of measurement, the material is traditionally cut into four-foot lengths and stacked into piles known as cords. Two stout stakes driven home either end keep the stacks in place.

A cord has fixed dimensions. The standard size is four feet by four feet by eight feet (1.21 metres x 1.21 m x 2.44 m), but regional variations exist. This gives a true volume of 128 cubic feet (about 3.6 cubic metres), but the actual volume of wood is about 70 to 90 cubic feet (about two to three cubic metres) depending on straightness. A well-grown oak tree containing about two to three cubic metres of timber can be expected to produce about one cord of branchwood.

A cord of freshly felled material weighs about 1.5 tonnes, while a cord of seasoned material weighs about one tonne. Apart from firewood, the material is used for the production of charcoal and other uses where straightness and quality is unimportant.

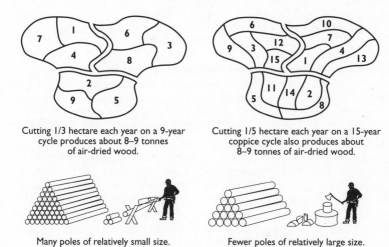

Cutting 1/3 hectare each year on a 9-year cycle produces about 8–9 tonnes of air-dried wood.

Cutting 1/5 hectare each year on a 15-year coppice cycle also produces about 8–9 tonnes of air-dried wood.

Many poles of relatively small size. Easy conversion with bowsaw.

Fewer poles of relatively large size. Logs may need splitting.

Figure 2.10 *Firewood Coppicing: Examples of Cutting Sequence in a Three-Hectare, Fully Stocked Wood*

required to produce one tonne of air-dried timber). About three hectares will be needed, therefore, to produce eight to nine tonnes of air-dried wood each year. (It is possible to double the output with certain willows, poplars, alders, southern beech and eucalyptus, but it will generally be found easier to grow these as new copses rather than attempting to convert existing woods.)

The interval between cutting the coppice should be fixed to yield material convenient for manual handling and conversion. A rotation of six to 15 years should produce stems in the range six to 12 metres tall and seven to 15 centimetres in diameter. To estimate the area of annual cut (or coupe), the total area to be worked is divided by the number of years in the coppice rotation. For example, if three hectares of evenly stocked woodland are to be coppiced on a nine-year rotation, each coupe will occupy one third of a hectare.

A mixed broadleaved copse worked for firewood on a nine-year cycle should have a stocking of about 1000 stools per hectare. Production of suitable sized poles will be achieved a year or two earlier if the number of shoots on each stool is reduced to about five in the second year after cutting. This concentrates growth on a smaller number of shoots.

Depending on the nature of the existing woodland, the first round of fellings may yield irregular quantities of timber from one year to the next, and firewood deficiencies may have to be met until the whole area is brought into management.

Preparing the Firewood

Burning green wood is inefficient. It is more difficult to light, yields less heat and causes tar deposits in stove and flue. The moisture content needs to be about 20 per cent or less. This can be achieved by stacking the wood on a sunny site until autumn, then cutting it into suitable sizes and storing it in a shed or other dry place with good air circulation. At least six months should be allowed between felling and burning, though for species such as elm, a year is necessary. If sufficient storage space is available the ideal system is to store for three years – with three-year-old firewood in current use, two-year-old material for use as a standby in periods of heavy demand and the present season's wood on hold. The amount of space needed per tonne of air-dry wood will depend on wood density. As a guide, beech, hornbeam and oak occupy about 1.5 cubic metres of space per tonne.

A range of firewood processing equipment is available, including benches incorporating a power take-off (PTO)-driven circular saw and usually a ram or cone-screw splitting device.[21]

Suitable Species for Open Fires

Table 2.5 lists some common broadleaved species ranked in an approximate order of usefulness for burning on open fires. Due to their resinous properties, conifers are not generally recommended for open fires. Most species are suitable for burning in closed stoves.

WOODCHIPS

The conversion into woodchips of otherwise unprofitable timber and large branches can produce a saleable product. As well as delivered-in logs, some chipboard manufacturers will accept woodchips produced in the wood, but the price offered limits the viable haulage distance.[22] Before sale, chips must be stored in well-ventilated areas and the heaps must be turned regularly to reduce the possibility of overheating.

Woodchip-fired heating systems hold potential for homes, outbuildings, greenhouses and swimming pools – wherever a regular and sustainable supply of wood can be guaranteed. Woodchips are also used for:

■ surfacing indoor equestrian establishments;
■ outdoor horse gallops;

Table 2.5 *Value of Various Species as Firewood*

Species	Value as Firewood
Ash	The most sought-after firewood. Rapid burning even when freshly felled. Emits a great amount of heat. Particularly valuable for fire-lighting.
Field Maple	Comes second in the list of useful fuels.
Beech	One of the best firewoods but difficult to light. Best stored for a year.
Birch	Burns freely and emits considerable heat.
Elm	Makes a pleasant and long-lasting fire when partially seasoned.
Hawthorn	One of the best firewoods for lasting quality and amount of heat emitted, either green or seasoned. Inclined to spark.
Hornbeam	One of the most valuable timbers. Emits a great amount of heat.
Hazel	Well known for its value as firewood. Much sought after.
Oak	Hard to beat for lasting properties and amount of heat produced but difficult to light.
Sycamore	Considerable heating properties.
Apple	Good, with pleasant smell.
Alder	A reasonable firewood.
Wild Cherry	Difficult to start but once alight lasts long and gives a fierce heat. Best when old and partially seasoned.
Holly	Very good, even when green.
Horse Chestnut	Burns reasonably well but spits. Best stored for a year.
Sweet Chestnut	Not the best firewood. Inclined to smoulder but can spark viciously. Must be well seasoned.
Lime	Difficult to light and third rate as firewood. Inclined to smoulder. Difficult to split.
Willow	Wanting in heat. Inclined to smoulder.
Poplar	Emits only a small amount of heat. Inclined to smoulder and produces acrid smoke.
Elder	Of little value. Very acrid smoke.
Conifers	Generally inclined to spark with no great heat value. Best in wood burning stoves.

- footpaths;
- horticultural growing mediums;
- soil conditioners;
- mulches;
- fuel for electricity stations.

CHARCOAL

Charcoal is made when wood is heated in the almost complete absence of air. Converting wood into charcoal doubles its energy value and considerably increases its commercial value. About six to seven tonnes of wood are needed to produce one tonne of charcoal.[23]

The production of charcoal is one of the oldest of country crafts, with evidence of its use going back thousands of years. The traditional method involved the preparation, on the woodland floor, of a level circular hearth, five to six metres in diameter, on which a huge stack, or kiln, of wood was built using short lengths of seasoned wood. The kiln would be built with a hollow centre that served as a flue and the whole was finished off with a layer of earth or grass turves. Red-hot charcoal from a previous 'burn' would then be dropped down the flue and when combustion was well underway the central hole would be capped. The whole process could take a couple of weeks and, depending on its size, one kiln might produce ten cubic metres of charcoal.

The men who made the charcoal were out in all weathers cutting and stacking wood in the autumn and winter, when the sap was down, and burning during the rest of the year. Hardy people, they lived a simple life in the woods – often with their families – in cramped conditions in turf-covered huts or in caravans. The kilns had to be carefully watched night and day for signs of weakness in the earth layers that might allow the fire to flare, destroying days of work in moments. The traditional form of charcoal production is now seen only in demonstrations.

Market Potential

An estimated 60,000 tonnes of charcoal is currently imported annually, mostly for barbecues.[24] It is also used in the smelting of high-quality steel, the refining of non-ferrous metals and horticultural composts. To produce 60,000 tonnes requires the conversion of almost half a million tonnes of wood, yet UK production in 1995 accounted for less than a 5 per cent share of the market. Production is steadily increasing, however; from very small numbers a few years ago there are now several hundred charcoal burners in Britain, most using modern, portable kilns. Though sometimes a little more expensive than the imported material, home-produced barbecue charcoal is considered to be of a higher quality. It lights more readily, comes to cooking temperature quickly and burns cleanly. The big sales advantage for the retailer is that, coming from properly managed British woodlands, charcoal has a 'green' profile that appeals to many people.

Small woods in need of restoration and young broadleaved plantations in need of thinning hold enormous quantities of low-quality material suitable for few other markets. Charcoal could also provide a useful outlet for material from tree surgery work. Most broadleaved species are suitable for charcoal production. Beech, hornbeam, ash, oak, elm and hazel are the preferred species, but sweet chestnut, birch and sycamore are also suitable. Poplar is generally considered too soft. Conifer charcoal is very friable, though it lights quite easily.

Equipment

A modern kiln consists of a drum of one or more circular, stack-able, interlocking steel rings of about two metres diameter and almost the same height. Each unit incorporates a lid, a set of tubular chimneys and air controls. A single-ring kiln can produce about 270 kilogrammes of charcoal in a single burn lasting up to three days. A two-tier kiln can produce about 400 kilogrammes of barbecue-grade charcoal from about 3.25 tonnes of green timber.[25] Where wood supplies are plentiful, two kilns can be employed for greater efficiency, one being loaded or unloaded while the other burns or cools. The finished product has to be sieved to remove the dust and is then graded, weighed and bagged.

A high level of expertise is required to obtain consistently good results. The charcoal burner must be constantly on hand to monitor progress and to adjust the air flow as necessary. Would-be investors in charcoaling equipment should consider the following:

- Making charcoal is a skill that calls for proper training.
- The work is one of the toughest and dirtiest in the country-side.
- To be viable, the venture requires large reserves of wood.
- Before use the wood must be converted, stacked and left to season for six to 12 months.
- The operation can produce a lot of smoke, so the effect on neighbours should be considered.
- VAT is currently charged at 8 per cent if charcoal is used as a fuel and at 17.5 per cent if sold for industrial use or for agriculture or horticulture.
- The production of charcoal for personal use gets the full added value.

Box 2.11
HOW TO MAKE SMALL QUANTITIES
OF CHARCOAL

A clean disused 40-gallon drum can be used to make charcoal. One end is completely removed and is fashioned to make a lid. In the other end half a dozen slits are punched. The finished kiln is placed on a few bricks to allow air to flow upwards through the holes and a fire is made in the bottom. Seasoned wood is then loaded into the kiln and when well alight the air is restricted from all but a small section around the base of the kiln by mounding up earth or turf. The lid is then placed on top, leaving a small gap for smoke to exit. When the colour of the smoke changes from white to blue all air is excluded by closing off the bottom vent and sealing any gaps in the lid with earth or turf. One burn will produce about 20 kilogrammes of charcoal.

Woodland owners should ensure that prospective operators have the appropriate equipment and skills to do the work properly, with minimum disturbance to wildlife, and will leave the woodland in a good condition. It is worth seeking the advice of the local authority or the Forestry Authority to ascertain whether a charcoal marketing company operates in the region.

Courses and seminars covering coppicing, coppice restoration, greenwood working, turnery, furniture making and charcoal making are held by various regional and national woodland initiatives on an ad hoc basis.

CHRISTMAS TREES

On paper, growing Christmas trees can seem a very attractive proposition, but their management calls for skill and commitment. Christmas trees are generally grown on a rotation of seven to nine years. Establishment will be found easiest where the trees are planted on weed-free, ex-agricultural land, but there may be convenient areas within existing woods that can be utilized. Planting under power lines is a possibility but it may require the prior agreement of the appropriate authority. Christmas trees may also be grown as a 'nurse' crop between broadleaved species; however, wherever they are grown, they must be sited where there is minimal chance of theft.

Trees must be fenced against rabbits, hares and deer, and timely weeding is essential. Weeds will reduce both growth and survival rates and will smother lower branches, leading to poor-quality trees. Great care must be exercised if herbicides are to be used – years of work can be wiped out in minutes by one inexpert application.

The traditional Christmas tree is the Norway spruce, but a number of other species (such as Scots pine and silver fir) are also popular. These tend to be slower growing but they command higher prices. Most species are tolerant of a wide range of sites but dry, alkaline and frosty sites are best avoided. The trees are usually planted one to 1.2 metres apart, giving a planting density of 7000 to 10,000 trees per hectare. The proportion of first-quality trees can be increased by 'shearing'. This involves pruning and shaping the leading shoot and side branches to give a denser, more compact form. Secateurs or a motorized pruning saw are used. Correct timing and accurate cutting are critical factors. Expert advice should be sought before shearing.

Most trees are cut or lifted when they are 1.2 to 1.8 metres in height. They must be harvested between mid-November and mid-December to ensure they reach the customer in the best possible condition. At this time of the year it is also possible to utilize selected tops of conifers felled for other markets. Further information on all aspects of the subject may be obtained from the British Christmas Tree Growers Association (see list of addresses in Appendix 4).

POPLARS

Many landowners planted poplars in the 1950s and 1960s as a result of encouragement from the match manufacturing industry, but a single market, a dying enterprise and a tendency for poplars to develop bacterial canker thwarted most expectations.[26] In the late 1970s the British market for poplar timber for matches collapsed, and as a result thousands of plantations now stand unmanaged and their owners are at a loss to know how best to market the trees. The timber is seldom of high quality and rarely commands high prices, though it can still attract buyers.

A revival of interest and support for poplar cultivation is now apparent as a result of trials of new Belgian poplar 'clones' that are less prone to bacterial canker and major leaf diseases.[27] Some can achieve an incredible four metres in the first growing season and can produce trees measuring 50 centimetres dbh at 12 years old.[28]

Poplars are usually planted as unrooted sets (entire shoots) or cuttings (parts of shoots). Sets should be pit-planted using a spade for best results, but provided that the soil is friable and well culti-vated they can be inserted into a hole made with an iron bar. They should be planted with one third of their length below the soil surface. Cuttings can usually be pushed into well-cultivated soil by hand. All but the top two to five centimetres of a 25-centimetre cutting should be below the soil surface.[29]

Poplars are exacting species. They require a fertile, moist (but not waterlogged) soil that allows deep rooting. They do well on the banks of watercourses but they must be adequately weeded. They are also light-demanding trees. Traditionally, they were planted at very wide spacing (for example at eight metres by eight metres) and there was never any intention to thin them out. Today, some growers plant the new varieties as close as two metres by two metres on rotations of around ten years to produce high yields of pulp or fibre. Other growers are intent on longer rotations, with the trees being thinned for firewood or other sales and the best being grown on for timber. The trees destined for final crop should be regularly pruned if the aim is to produce high-quality stems suitable for veneers.

CRICKET BAT WILLOWS

Cultivation of the cricket bat willow (*Salix alba* var *caerulea*) is not difficult but close attention to detail is necessary, especially during early life.[30] Bat willows are best grown near running water. Badly drained sites are unsuitable. Valley bottom sites with alluvial soils over sand are ideal. Occasional flooding during the growing season does not matter, but marshy sites with rushes and sedges indicate that conditions are too wet. The tree is not suited to windswept, humid or coastal areas. Southern England appears to be the region best suited to its needs.

Unrooted cuttings up to four metres in length are planted in late winter at a spacing of eight to ten metres apart into deep holes made with an iron bar. Pouring a little water in the hole makes for easier work. Bat willows are often planted as single rows alongside streams.

During early life there is a tendency for buds to develop on the stem. To produce knot-free timber these buds should be rubbed off by hand as high as can be reached. This should be done two or three times each season for as long as new buds appear. No thinning is required and the trees should be ready for felling at 15 to 25 years, depending on the rate of growth. They can usually be sold to a cricket bat manufacturer who will arrange for them to be expertly felled and new trees planted in their place.

An important bacterial condition (watermark disease) affects the tree. It is widespread in south-east England and the only method of control is the felling and destruction of infected material.

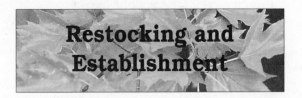

Restocking and Establishment

PREPARING THE GROUND

Restocking should follow on as soon as possible after felling (within days rather than months). Provided the harvesting operation is carefully planned and properly executed, the felling area should be left in a reasonable condition ready for restocking. It is a good plan to have the same team do the felling and the restocking (though good 'fellers' do not always make the best planters) to help ensure that the need for further preparatory work is kept to a minimum. However, several factors can retard progress.

Surface Water

Harvesting operations can leave the felling area in a rutted and waterlogged condition – particularly when the weather has been wet, where the site is level, or where soils are clayey or peaty in texture. A rise in the watertable is normal following the removal of a tree crop but can be a cause of subsequent plant losses. Drains may have to be cut to carry surplus water off the site.

Lop and Top

Smaller branches and the tops of felled trees are generally left where they fall. Under normal circumstances this 'lop and top' will slowly decay and will cause no problems. It can provide valuable wildlife habitats as well as a source of nutrients for the next tree crop. It may also help restrict the movement of deer on the restocked site. In excess, however, it can impede access, render planting and weeding more difficult and harbour rabbits. The decision on what constitutes a reasonable quantity to leave will depend on individual circumstances. It can be gathered into heaps, by hand, to be burnt on site, though great care is necessary to ensure that fires do not get out of hand, putting the woodland at risk. No fire should be built so close to retained trees that they are likely to be scorched, and all fires must be properly turned in before the end of the working day.

Ground Vegetation

Where a presence of dense vegetation threatens to compete with the new plants, an appropriate herbicide may be applied. Broadcast application should be avoided wherever possible – it is generally considered environmentally unacceptable. Spot-application of one-metre diameter circles at the proposed plant spacing is a better method.

Compacted Ground

Every effort should be made to restrict the movement of heavy machinery on sites that are to be replanted. It is a mistake to use a tractor to heap brash or to remove surface vegetation. Any compacted ground should be adequately cultivated before restocking, otherwise trees will not thrive. This may not be possible where stumps and roots are numerous.

EXPLOITING REGROWTH FROM THE PREVIOUS CROP

Planting is not the only way to restock a felled site. It may be possible to use regrowth (coppice, natural regeneration or root-suckers) from the previous crop, but to be successful it must be adequately protected from browsing animals.

Coppice

Any regrowth from the stumps of the felled trees may be recruited to form all, or part, of the next crop. It can be allowed to develop as a crop of coppice with each stump carrying a large number of stems, or the regrowth can eventually be singled to one stem per stump and allowed to develop as normal woodland. Such trees do not usually produce the highest-quality timber.

Natural Regeneration

The woodland floor often holds a rich deposit of tree seed that will germinate in the first season or two after felling, due to increased light and warmth. In woods where nature conservation is important, the recruitment of natural seedlings may be more desirable than planting. Natural regeneration conserves local genetic features, creates more diversity and produces a wide choice of stems for selection.[1] Regeneration of hawthorn, elder, birch, field maple and other species can help soften formal woodland edges and will become a valuable wildlife resource. On the other hand, the growth of pioneer species, particularly sycamore, may compete with, and suppress, planted trees.

As a management system aimed at restocking woodlands without the need for planting, natural regeneration has never been widely used in Britain. It requires intensive management, it is unpredictable and woodland animals are inclined to predate on the seed. There are no generalized prescriptions for success and it is rarely a cheap or an easy option. It is most likely to succeed where:

■ woods are well stocked and regularly thinned;
■ dominant trees are evenly distributed;

- seedlings are already present;
- selective felling follows the appearance of good regeneration;
- predatory and browsing animals are scarce or properly controlled;
- growth of weeds is poor or adequately controlled;
- treeshelters are used to aid establishment;
- soils are free draining and infertile.

Sites with poorly drained soils allow rapid growth of weeds and are likely to be particularly problematic.[2]

Root Suckers

Some trees (for example, wild cherry, aspen, elm and poplar) can reproduce by means of sucker growth that arises from the roots of standing or recently felled trees. The effect is to form a 'clone' of trees that are all genetically linked to the parent tree or trees. Such groups can be managed as a part of the growing crop.

Regrowth, of whatever sort, is not always welcome. Where it is not wanted it may have to be cut back several times until the new crop begins to suppress it. Grubbing-out the stumps is usually impractical and environmentally undesirable. Unwanted coppice regrowth can be controlled before it arises by chemically treating the freshly felled stumps, or later by spraying the first season's foliage.

PLANTING

Planting is the most usual method of restocking felled sites. Depending on circumstances and the owner's objectives, it is advisable to:

- aim for informality in the layout;
- avoid straight rows and geometric patterns – they make for greater efficiency but impair visual amenity;
- where straight rows are essential, plant along the contours and vary the distance of plants in the rows;
- plant large 'forest-type' species towards the centre and back of the scheme as seen from the most important viewpoints;
- plant trees of smaller stature at the edges and in the foreground;
- leave parts unplanted to become wildlife glades;
- redesign straight rides to incorporate sweeping curves;
- incorporate some shallow bays in rides to produce a 'scalloped' effect;

- ensure rides are sufficiently wide – make due allowance for the growth of side branches;
- consider the next round of felling – plan the location of extraction routes, work areas and loading bays.

Choosing the Species

The selection of species is probably the single most important decision for success in woodland management; no matter how carefully trees are planted and maintained, if the selection of species is unsound the final result will be disappointing and will endure for years. Choice of species depends on:

- the owner's objectives;
- the potential markets for produce;
- the site characteristics;
- any existing or perceived management constraints;
- the availability of planting stock.

A wide range of objectives can be achieved by careful selection of species. Different objectives can usually be brought together in one planting scheme and only occasionally will compromise be necessary. In making the decision:

- Consider whether the previous crop was satisfactory – there may be no need for a change.
- Note which trees grow well on similar sites in the locality – but bear in mind that some species can grow quite well when young, only to fail in middle age.
- Remember some species are light-demanding; do not plant beneath the crowns of retained trees or in very small clearings.
- Consider willow, poplar or alder for wet sites, perhaps planting on mounds to raise roots above the watertable.
- In very wet conditions, consider leaving the area unplanted to develop as a wetland habitat or excavate a pond for wildlife.
- Consider using conifers as a nurse amongst broadleaves to stimulate height growth and to create shelter – but aim to remove them before they interfere with the development of main crop trees.
- Where blocks of conifers are to be planted, retain a variety of native trees, shrubs and ground flora.
- Where grey squirrels are numerous, avoid beech and sycamore – ash and cherry are safer choices.

- Do not use more than about 10 per cent of wild cherry in any planting scheme: they are at risk of certain root and bacterial canker disorders.
- Use elm (and wych elm) very sparingly, if at all – they will almost certainly develop Dutch elm disease.
- Remember the foliage of yew is poisonous to farm livestock.
- Resist any impulse to plant a large number of species – it can look unnatural.
- Avoid too exotic a choice of species – false acacia, Japanese flowering cherry and laburnum are beautiful trees in their own right but their use in woodland situations is questionable.

When sites that have been group felled are to be replanted, care is needed to keep light-demanding species well away from the shade cast by surrounding trees. On no account should they be planted directly under the crowns of existing trees. When dealing with deciduous trees, remember that what might appear a fair-sized clearing in winter may shrink to a tiny gap in the canopy when the trees are in full leaf.

Native Species

Only about a dozen trees played a principle role in shaping our native woods, with just one or two species in the uplands. For small schemes, broadleaved trees and shrubs that are native to this country, and in particular to the locality, will often be the choice. They can satisfy most objectives and create a pleasing result. Furthermore, using plants from local sources helps to preserve biodiversity. Native trees have adapted to the huge geographical and climatic differences that exist in Britain. Scientists point to the damage that can be caused to local flora and fauna when plants from outside Britain, with differing flowering and fruiting times, are used. For areas identified as ancient woodland only native trees should be used (see Appendix 1). A number of tree nurseries are now specializing in plants raised from home-collected seed sources.

Mixtures

Mixtures generally allow more flexibility in meeting market demands and may improve overall profitability, but they require a higher degree of attention and skill so it is wise to keep things as

Table 3.1 *Tolerance of Trees to Shade*

Light-Demanders	Moderate Shade Bearers Broadleaves	Shade Bearers
Ash	Ash (when young)	Beech
Oak	Sycamore	Hornbeam
Sweet chestnut	Lime	
Poplars	Hazel	
Birch		
Alder		
Willows		
	Conifers	
Scots pine	Coast redwood	Western red cedar
Corsican pine	Douglas fir	Western hemlock
Larches	Lawson's cypress	Yew
		Noble fir

simple as possible. Some species are compatible and will grow well together, even when planted in an intimate mixture. Oak, ash and wild cherry will do this. Other mixtures may result in vastly different rates of growth, with the possibility of slower-growing species becoming suppressed and dying out.

One way of helping to ensure that all the species in a mixture are represented throughout the life of the crop is to plant single-species groups or drifts containing at least a dozen plants. The traditional forestry practice of planting rows of alternating species and other geometric patterns rarely finds favour today.

The Woodland Edge

The design of that part of a restocking area, where it meets the boundary of a wood or the edge of a ride, requires special attention. The majority of woodland animals, birds and insects are to be found in these edges, not deeper within the wood, and tree and shrub species that attract wildlife should be favoured here. If occasional patches of land are to be left unplanted to develop as wildlife glades, these are ideally sited in this zone. Species should be chosen to create a sloping woodland edge to eventually give a natural layered or 'cascade' effect. This will help to deflect the wind and create warmth and stability within.

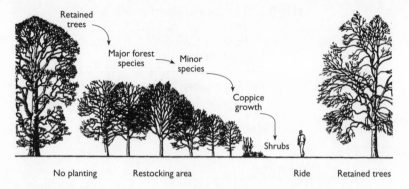

Retained trees

Major forest species

Minor species

Coppice growth

Shrubs

No planting   Restocking area   Ride   Retained trees

Figure 3.1 *Design of the Planting Edge Showing Cascade Effect*

Table 3.2 is a novice's choice of species guide intended for small areas of restocking in the order of one hectare or less. It may also be useful for larger areas where management objectives are not demanding. For other situations, and particularly where the production of timber is paramount, professional advice should be obtained.

Spacing

Where the primary objective is to produce quality trees for timber production, a close spacing helps to prevent the development of coarse side branches and also provides larger numbers from which to select the best individuals. Trees planted at wide spacing must be judiciously pruned if they are to produce good timber.

Where environmental objectives are important, wider spacing will permit the recruitment into the planted crop of any useful natural regeneration, coppice or sucker growth, thus adding diversity.

Types of Plants

Under European directives, certificates of origin are a legal requirement for oak, beech, poplar and the main commercial conifers that are to be grown for timber production. A number of forest nurseries now employ seed collectors to ensure the supply of genuinely native stock for other tree and shrub species and for those grown for other objectives. The type and size of plant will depend on the site, the species to be used and the finances avail-

---

Box 3.1
HOW TO USE THE CHOICE OF SPECIES ORGANIZER

1 Select the main management objective in Table 3.2 and tick all corresponding boxes in the centre column.
2 Identify any relevant site constraints and cancel any relevant ticks in centre column.
3 Noting that some boxes in the centre column are shaded, compile a list for replanting as follows:
   ■ unshaded boxes = these are the main species; select no more than three, to be allocated to 50 to 60 per cent of the planting area;
   ■ light shaded boxes = these are the minor species; select about three (no more than four), to comprise 20 to 30 per cent of the area;
   ■ dark shaded boxes = these are for occasional use on boundaries, etc; use in small groups.
4 From the shrub list select as many as appropriate and allocate to about 10 per cent of the area. Use on edges, boundaries and along rides, or interplant in main planting area to form an understorey.
5 If any conifers are desired it is best to seek professional advice before proceeding; many are not compatible in close mixture with broadleaves.
6 Finally, allow for 10 to 20 per cent of the area to be left unplanted to be managed as wildlife glades or other open spaces.

---

able. Provided they are sturdy, small plants often survive better than larger ones.

Transplants are the most usual tree type for replanting in woodlands. They are cheaper to buy and to plant than taller trees and do not require staking. Containerized plants are less prone to handling damage and do not have their roots exposed during planting, so they suffer little planting shock. On the other hand, they are more expensive, require more storage space and, where the ground being planted is very different in character from the growing medium, the roots may have difficulty in adapting.

Time of Planting

Container-grown plants may be planted throughout the year in appropriate weather conditions. Traditionally, the planting season for bare-rooted (non-containerized) stock is during the dormant period, from late October to the end of March. Autumn planting is

Table 3.2 *Replanting in Small Woods:
Choice of Species Organizer*

Species	Landscape	Wildlife	Timber Production	Firewood Production	Sporting	Shelter	Screening	Very Dry Site	Very Moist Site	Frosty Site	Sandy Soil	Heavy Clay Soil	Calcareous Soil	Very Exposed
Broadleaves														
Alder, common		✔					✔	X						
Ash	✔	✔	✔	✔	✔		✔			X				
Aspen	✔	✔		✔			✔	X						
Beech	✔		✔			✔	✔		X	X				
Birch	✔	✔		✔	✔	✔	✔	X						
Chestnut, horse	✔						✔		X	X				X
Chestnut, sweet	✔		✔			✔	✔		X	X		X	X	X
Hornbeam			✔		✔	✔								
Lime, small-leaved	✔	✔		✔			✔	X	X					X
Maple, Field	✔	✔							X					X
Maple, Norway			✔						X	X				
Oak	✔	✔	✔	✔	✔	✔	✔	X		X	X			
Oak, Holm						✔	✔	X	X			X	X	
Poplar, Grey	✔	✔						X			X			X
Poplar, hybrids	✔		✔	✔	✔		✔	X			X			X
Sallow		✔						X			X	X		
Sycamore			✔	✔		✔	✔	X	X		X			
Walnut			✔					X	X	X		X		X
Wild Cherry	✔	✔	✔	✔	✔			X	X		X			X
Wild Service		✔						X	X	X	X			X
Willows	✔	✔					✔	X						
Conifers														
Douglas fir			✔					X				X	X	X
Larch, European	✔		✔		✔			X				X	X	X
Larch, Japanese			✔		✔			X				X	X	
Pine, Corsican			✔			✔	✔		X					
Pine, Scots	✔		✔			✔	✔		X				X	
Spruce, Norway			✔				✔	X				X	X	X
Spruce, Sitka			✔				✔	X				X	X	
Western Red Cedar			✔				✔		X					
Yew	✔						✔		X		X			

Shrubs

Species	C1	C2	C3		C4	C5	C6	C7	C8	C9
Blackthorn	✔		✔							✗
Buckthorn, Alder	✔				✗					✗
Buckthorn, Purging	✔				✗					✗
Crab apple	✔	✔								✗
Dogwood	✔	✔				✗				✗
Guelder rose	✔	✔			✗					✗
Hawthorn	✔	✔	✔			✗				
Hazel	✔	✔	✔			✗				
Holly	✔	✔	✔			✗		✗		
Privet		✔	✔			✗				
Rowan	✔	✔				✗	✗	✗	✗	
Spindle	✔				✗	✗				✗
Wayfaring tree	✔					✗				✗
Whitebeam	✔					✗	✗			

Notes: Instructions for use are given in Box 3.1 on page 111. Native species printed in bold type.
Source: Ken Broad, 1994

generally regarded as more successful. Late spring plantings are more susceptible to early summer droughts. The planting season for evergreen trees is slightly longer than for deciduous trees – from late September to early April. Other points to note:

- Planting is best undertaken when the soil is moist and friable – do not plant into frozen or waterlogged ground.

Table 3.3 *Approximate Number of Trees per Hectare for Various Spacings*

Spacing (m)	Spacing (ft)	Number (ha)
0.9 × 0.9	3.0 × 3.0	12,350
1.2 × 1.2	4.0 × 4.0	6950
1.5 × 1.5	5.0 × 5.0	4450
1.8 × 1.8	6.0 × 6.0	3100
2.0 × 2.0	6.5 × 6.5	2500
2.1 × 2.1	7.0 × 7.0	2250
2.4 × 2.4	8.0 × 8.0	1750
2.5 × 2.5	8.5 × 8.5	1600
2.7 × 2.7	9.0 × 9.0	1350
3.0 × 3.0	10.0 × 10.0	1100
3.2 × 3.2	10.5 × 10.5	1000
3.4 × 3.4	11.0 × 11.0	850
3.5 × 3.5	11.5 × 11.5	800

Table 3.4 *Types of Planting Stock*

Type	Description
Standard	About three metres overall with two metres of clear stem, a well-balanced head of branches and an upright central leader.
Feathered tree	Stem well furnished with lateral shoots and with a straight upright leader.
Whip	Single stemmed tree, one or two metres.
Transplant	Sturdy tree with good fibrous rootstock usually 15 to 50 centimetres tall (20 to 40 centimetres common).
Containerized plant	12 centimetres to three metres tall, grown in a cellular tray or pot containing balanced compost.

■ Ensure replanting takes place soon after felling to reduce the need to deal with competing ground vegetation.

■ Order plants in good time – do not wait until the end of the growing season.

It is best to visit several tree nurseries to ensure suitable trees are obtained, and once good stock has been identified it is worth asking if it can be reserved until required.

Handling Plants

Moving young trees from nursery to planting site involves them in considerable stress. Survival can be affected by rough handling

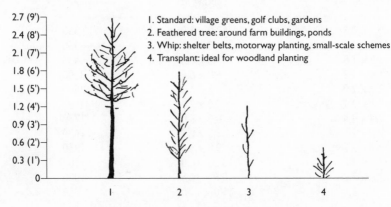

1. Standard: village greens, golf clubs, gardens
2. Feathered tree: around farm buildings, ponds
3. Whip: shelter belts, motorway planting, small-scale schemes
4. Transplant: ideal for woodland planting

Figure 3.2 *Types of Planting Stock*

Box 3.2
HOW TO HANDLE PLANTS

Ask the nursery how the plants will be packaged. If they are to be delivered in polythene bags they can be stored upright in their packaging for up to seven days, provided they are under cover, in cool conditions and away from sunlight. For longer periods the trees will need to be removed from the bags and heeled in (set out in prepared trenches, with the roots completely covered with well-cultivated soil). They should be spread out thinly to help prevent the onset of mildew. Bare-rooted plants should be heeled in immediately on receipt. The bundles should be untied and the plants spread out if long storage is unavoidable. Containerized plants can be safely stored for longer periods than bare-rooted trees.

Every effort should be made to have the trees delivered to the site as soon as possible after they have been lifted in the nursery and to have them planted as soon after delivery. Trees left for weeks in polythene bags, in draughty sheds, or even in properly prepared heels, will deteriorate.

Where conifers are to be planted there is a serious risk of damage by certain weevils and beetles. Specially treated plants can be purchased but gloves must be worn to avoid contact with the insecticide.

during transport or storage.[3] Many plant deaths can be attributed to poor handling techniques.

Planting the Trees

Planting is a skilled operation but with a little practice a novice can expect quite reasonable results. Several types of planting spade are in use in the forestry industry, but an ordinary garden spade will serve almost as well. There is rarely any need to add peat or any other planting medium to soils at the time of planting. On most lowland sites there is seldom need for any fertilizers.

Staking

Planting in woodland conditions rarely calls for the use of stakes (other than those used to support treeshelters). Newly planted trees up to about 50 centimetres in height, whether container grown or bare-rooted, should have stems that are able to hold their branches without the need for support. As the trees grow they adapt to the prevailing conditions. An unsupported tree swaying in the wind will develop an increased stem diameter over one that is staked.

1. Prepare trench with one sloping side

2. Place bundles in trench; replace soil and firm around roots

Figure 3.3 *Heeling In*

Stakes are an additional expense. Where their use is justified (for example, where standard trees are to be planted), it is important that the tree is not attached too high up, and that the tie is not left on too long. A stake extending to about one third the height of the tree is adequate, with a single flexible or adjustable tie.[4] The ties need to be inspected and loosened from time to time. In many cases the supports can be removed at the start of the second growing season after planting to provide an opportunity for the trees to achieve a natural balance. Sudden removal of a high support after many years can result in stem fracture in strong winds. It is essential in these circumstances to gradually reduce the height of the tie over several years.

Box 3.3
HOW TO PLANT TREES

The planter should be equipped with a spade and a planting bag (old fertilizer bags should be avoided – they may contain residues that can damage plants). Plants should be removed from the polythene bag or taken from the heel, one bundle at a time, and placed in the planting bag.

Remove weeds by 'screefing' with the spade and dig up any brambles and bracken from the planting position. For standards, whips, shrubs and containerized plants, dig a hole of sufficient size to accommodate the roots. Break up any subsoil in the bottom of the pit. For smaller transplants it may not be necessary to dig a hole – a simple notch is all that is usually required.

Take the plants out of the planting bag one at a time and plant without delay. Do not expose them to the elements longer than is absolutely essential; roots can suffer irreversible damage if they are allowed to dry out. Plant the tree to the same depth as it was grown in the nursery – look for the distinctive 'tide mark' on the root collar. Replace the soil and lightly firm, or close up the planting slot using the heel of the boot. To encourage a well-balanced root system, avoid planting too close to old stumps.

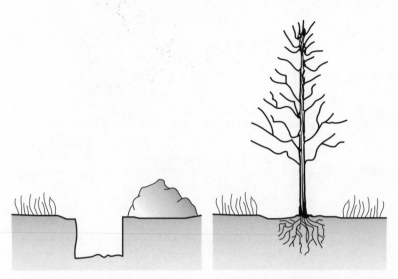

1. Dig hole large enough for roots; cultivate bottom and sides

2. Insert plant, backfill and lightly firm to exclude air

Figure 3.4 *Pit-planting*

Watering

Watering trees after planting or in drought conditions is rarely a practical proposition. It is easier and cheaper to accept some losses and replace failures.

Fertilizing

Only on the poorest soils (such as deep peats) should fertilizers be considered at the time of planting. The typical range of woodland soils (certainly those found in most lowland situations) do not require fertilizers.

Mulching

Mulches can be used as a practical and environmentally friendly alternative to chemical weed control. An effective mulch kills weeds, prevents weed seeds from germinating and conserves moisture. It can take several forms and utilizes either synthetic or organic materials.[5]

117

1. Screef away surface vegetation; insert spade and rock back and forward to form notch

2. Insert roots well down in hole

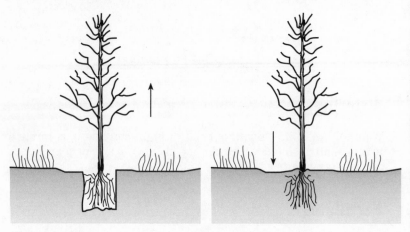

3. Pull plant upwards a little to straighten roots

4. Lightly firm soil with heel to exclude air

Figure 3.5 *Notch-planting*

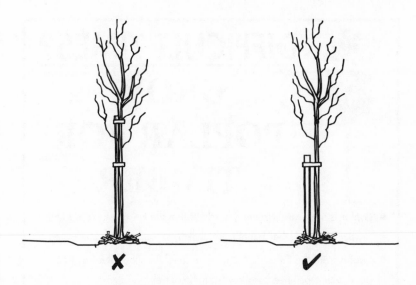

Figure 3.6 *Staking a Standard Tree*

Mulch Mats

These are manufactured as sheets of polythene, polypropylene or bitumen (but can also be custom-made from a roll of polythene) and are about one-metre square with a small central hole for the tree. They are secured using various types of pins and pegs or by using turves. Whatever the method it is important to prevent entry by voles. Where a treeshelter and a mulch mat are used together, the shelter must be erected over the mat, not the sheet around the shelter.

Treespats

These are similar to mulch mats but have a single slit from the centre to one edge, allowing it to be overlapped.

Bark and Woodchips

These can be laid some 50 to 70 millimetres thick in a circle of about one-metre diameter around the tree. They need to be properly composted before use so as to remove any toxic substances. While they stop the growth and germination of weeds, any seeds landing on the surface may still germinate.

A NEW OPPORTUNITY
FOR POPLAR

*The specialist potential
for Poplar by Hugh Snell*

It can be fun growing Poplar for timber, and a satisfactory visual effect can be achieved very quickly when Poplar are close planted and managed correctly. Put in perspective, it is perfectly possible to produce a crop of thinning before year 10 and a final crop of timber by year 25. This commercial return would be twice as fast as that expected from a softwood crop. Careful, management is required however; but given enthusiasm any forestry owner should easily cope!

Poplar is of course a Hardwood. This chapter addresses growing Poplar for timber and does not consider its potential for biomass or short rotation coppice. Poplar is now a very important lowland crop, having fallen out of favour several decades ago following the decline of the British match industry, the species is attracting considerable interest all over the world. The reasons for this revival are twofold. Firstly, Poplar has benefited from a dedicated breeding programme in Belgium and elsewhere, and consequently there have been improvements in growth rate, also a careful selection process has produced disease resistance in addition to increased growth. Secondly a fresh approach has been made to the production and management of Poplar woodland. Together, these two factors offer Poplar growers a new incentive.

So what of the timber, what are its uses, and can it be sold? There are several rather unusual facts about Poplar. The wood doesn't splinter, and yet clefts quite as readily as ash or chestnut, it is light in weight and in colour, and the wood nails well and takes paints and preservative well too. Without treatment Poplar timber will rot beneath the ground, but above ground the wood survives quite as well as other woods. More and more uses are being found for Poplar. The wood is ideal for floorboards, picture frames, pallets and vegetable boxes. Good too for country kitchen furniture, which offers a special opportunity.

A reader considering planting any new woodland, for their own use for

121

firewood, or rails or wishing to produce veneer quality woods, should perhaps consider Poplar as an option, or include some in the plantation. A word of warning, young Poplar are very liable to attack by squirrels from as early as year four. Great care must be taken to prevent the bark peeling by squirrels, not easy to spot twelve or more feet from the ground.

Poplar grow well in most soil conditions, apart from shallow rocky soils or those which are waterlogged. Deep well drained sheltered soils are best. There is no need to plant rooted Poplar, unrooted Sets, any length from one metre to 4 meters should be chosen, and these should be planted with 25% beneath the ground, planting would normally take place during March or early April. By far the best results are obtained when the land is autumn ploughed and a frost tilth produced, a light cultivation should take place just prior to planting, and the Sets may be pushed into the ground very easily. A residual herbicide is best applied overall, and will not hurt the young plants if applied soon after planting. Subsequent management entails protecting the plants from rabbits, cheap 60mm spirals will suffice, periodic inter row cultivation and regular pruning, leaving only 30% of the top of the tree unpruned. Pruning can be dome at any time but July to September is probably best allowing rapid healing of the wound and minimum regrowth from the cut area.

Problems? - Canker is unlikely when registered stock is used. Rust attacks are possible dependant on the growing conditions. Rust is not normally

Close planted Poplar in their 1rst year.

PROTECTION

Young trees can be damaged by a large number of agents and protection can take many forms.

Fencing

Whether to fence a planting site or to protect individual trees depends on:

- the animals to be excluded;
- the management objectives;
- the degree of access required;
- the likely cost.

A properly maintained fence will protect trees for a longer period than individual tree protection, but fencing is generally cost-effective only where the planting area exceeds a couple of hectares. The exact area at which one is cheaper than the other depends on the shape of the area, the length of the boundary and the tree spacing. To minimize costs, fences should be kept as straight as possible and any trees outside the fence should be individually protected.

Fencing is essential to exclude farm livestock. It can be a disadvantage, however, where game birds need free access, and voles can multiply rapidly in the dense grass that develops after grazing animals are excluded. Individual protection may then be required in addition to a fence. If a heavy branch or a tree falls onto a fence the whole plantation may become vulnerable. A breach can go unnoticed for a long time. Drifting snow can also allow animals to cross a fence; on thawing, they may be fenced in rather than out.

The permanence of fences can also pose problems – a single boundary fence restricts the movement of ground-based wildlife, while if used to protect a series of individual felling coupes, such fences may impede access and thus hamper management activities.

There are several forms of fencing suitable for tree protection.

Wire-Mesh Fencing

For a fence to be effective against rabbits, the bottom 150 millimetres of the netting should be buried or turned outwards towards the rabbits and turfed over.

Traditional deer fencing is very expensive (though high tensile fencing can halve costs) and is rarely practicable for small schemes. It needs to be 1.5 metres high for roe and muntjac; 1.7

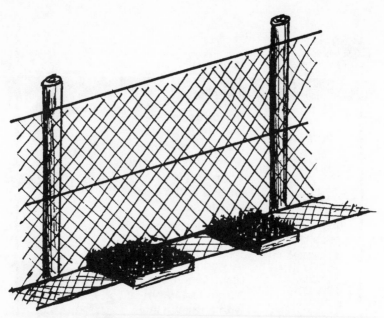

Bottom 150mm of netting turned out towards
rabbits and turves placed above

Figure 3.7 *Fencing Against Rabbits*

metres for fallow and sika; and 2.0 metres for red deer. Fences should never be erected across traditional deer 'runs' – this incites the deer to gain entry at any cost.

Electric Fencing

Electric fencing, often claimed to be a cheap form of protection, may not be an effective barrier against deer. Roe deer are deterred to some extent, but browsing damage and deer activity can still be found within electrically fenced areas.[6]

Dead hedges

These are hedges about 1.5 metres high and made of cut branches woven in and out of a line of stakes. They are a traditional and relatively inexpensive way of keeping fallow deer off recoppiced areas for a season but are of doubtful value against roe and will rarely be capable of keeping out muntjac deer. Dead hedges need regular maintenance – they can settle to less than half the original height after 12 to 18 months.[7]

Chestnut Paling Fences

Made from 1.5-metre high split chestnut palings stranded into wires about 75 millimetres apart, this type of fencing comes in rolls ready for attaching to stakes that should be spaced about four metres apart. It is quick to erect, does not look out of place and, although initially expensive, has a long life and can be moved to new areas as required.

Natural Barriers

Deer can be discouraged by the retention of some rough growth within and around the replanted area. Blackthorn is particularly effective for this purpose. Young coppice regrowth can be protected by building 'wig-wams' of branches or by laying branches carefully all over the site. Care is always needed to achieve a good balance between available daylight and the protection provided.

Treeshelters

Treeshelters are invaluable when planting small numbers of trees. For areas less than two to three hectares, the trees are generally more cheaply protected with treeshelters than with fencing. They are made from photo-degradable materials (polypropylene or PVC) and are designed to break down under the effect of ultra-violet light after about five years. Some models are designed to last seven to eight years to provide extended protection. Treeshelters are used to:

- protect trees from most woodland animals;
- speed the growth of most species;
- render trees easier to find in the weeding season;
- protect trees during chemical weeding operations;
- benefit plants in drought conditions.

Treeshelters will not compensate for poor choice of species, low-quality plants, rough handling, inadequate site preparation or lack of weeding. They come in a variety of shapes, heights and colours. The degree of protection is governed by shelter height.[8]

Treeshelters are generally supported with wooden stakes. The stakes need to be at least 25 millimetres square and must be firmly driven into the soil – deer like to rub against them and can push them over. Stakes last longer if cut from some durable timber such as oak or sweet chestnut, or if treated with preservative. Short treeshelters are often supported with bamboo canes – but bamboo

Tree shelters provide protection from predatory animals and herbicide applications, enhance growth and enable plants to be easily located

is prone to rotting and snapping after a couple of years in contact with the soil.

Other points worth noting:

■ On sloping ground taller treeshelters may be required.

Table 3.5 *Effectiveness of Various Heights of Treeshelters*

Animal	Shelter Height (m)
Rabbits	0.6
Hares	0.75
Sheep/roe and muntjac deer	1.2
Red, fallow and sika deer	1.8

- Whatever the height, treeshelters are not sturdy enough to protect against cattle and horses.
- Cone-shaped treeshelters are meant to be used wide-end uppermost.
- Any weeds that establish within treeshelters can grow rapidly and choke the tree – they must be removed manually.
- Opaque materials such as polyurethane pipe and old plastic fertilizer bags should never be used in place of treeshelters.

Some early treeshelters were made from a white material but the more usual colours seen today are less obtrusive browns and greens. A few of the earlier types have lasted too well and have not shown much sign of breakdown.[9] Where weeding is neglected, even modern treeshelters may not break down as quickly as desired due to the low incidence of ultra-violet light. This can result in the constriction of tree stems as they outgrow the treeshelters. Under these circumstances the treeshelters should be slit with a sharp knife before the stem fully occupies the diameter of the treeshelter. The treeshelter can then be left in place for a year or two to afford some continuing protection against mammal damage.

Box 3.4
HOW TO ERECT TREESHELTERS

Planting and sheltering are normally carried out in the same operation. Remove a screef of surface vegetation to leave a patch of bare soil. Plant a tree in the usual fashion and firmly drive in a stake alongside, ensuring the tree's roots are not damaged. Place a treeshelter carefully over the tree with the stake inside or outside the treeshelter, depending on the type. Press the treeshelter into the ground to ensure as close a fit as possible. Keep the top of the stake within about 0.4 metres of the top of the treeshelter. Two attachments are normally made to fasten or staple the treeshelter to the stake.

There appear to be few disadvantages in using treeshelters, although occasionally, particularly on the most exposed sites, they may not be the most appropriate method of tree establishment. The environmental transition that the trees experience on reaching the tops of the treeshelters may also lead to a sharp reduction in growth rate, shoot die-back and abrasion damage to stems where the trees eventually emerge.[10] Where treeshelters are used on old meadow land, ants can be a problem. They will sometimes build their nests inside the treeshelters and this can lead to the death of trees. Conditions within treeshelters can also allow aphids and other insect pests to increase rapidly, though this is seldom more than a local nuisance.

Recent developments in treeshelter design[11] have incorporated the following features:

- ratchet-type, releasable nylon ties;
- inbuilt perforation lines so that the tree can readily break through as it grows;
- treeshelters supplied with treated stakes ready to fit into an integral stake slot so that ties are not needed;
- treeshelters with strengthened sections to prevent ties from tearing through the material on windy sites.

There is some concern about the ultimate fate of all the plastic being widely used throughout the countryside. The material employed in most treeshelters is chemically unreactive and biologically inert, and is not considered harmful to animals or plants.[12]

Tree Guards

Tree guards are made from open-mesh high-density polyethylene or polypropylene. They are less expensive than treeshelters but they do not significantly influence growth rate, nor do they protect any branches that may grow out of the mesh. The material is available in rolls from which a wide range of heights and diameters can be cut. The material eventually becomes brittle, eliminating the need for removal. They are most commonly used to protect conifers, shrubs and beech trees (a species that does not always appear to benefit from treeshelters).

Spiral Guards

Spiral guards are opaque cylinders of plastic, spirally split from

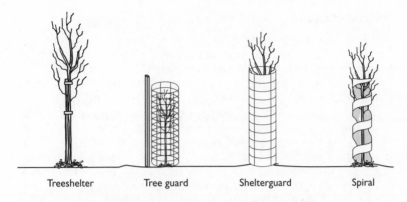

Figure 3.8 *Individual Tree Protection*

top to bottom. They are a convenient, low-cost means of protecting lower stems of trees against rabbits and hares, and are best used on whip-size trees at least 15 centimetres taller than the guard. They can weigh down small trees, so it is sensible to use stout canes as support. Small trees hidden within guards will almost certainly perish. Spirals expand with the growing stem. They are potentially reusable.

Chemical Repellents

Several proprietary mammal deterrent substances are available for protecting tree crops. They are said to give effective protection against rabbits, hares and deer, and may provide a more economical method of protecting trees than the use of fences, treeshelters or guards.[13] They can be useful in dealing with short-term problems while long-term protection strategies are worked out.

Chemical repellents are applied by dipping, spraying or painting vulnerable parts of the tree. The duration of protection depends on growing conditions. Fresh shoots will be vulnerable so over-winter protection is more effective.

Population Control

In situations where there is a risk of high levels of damage to trees by animals, direct control by shooting, gassing or a combination of methods may be necessary.

Grey Squirrels

The grey squirrel, a native of the eastern US, was introduced into Britain in the 19th century. It has spread widely, particularly in lowland areas with broadleaved and mixed woodlands where it is capable of attaining high population densities.[14] From mid-summer until the end of winter grey squirrels feed mainly on tree fruits such as hazel nuts, acorns, beech mast and conifer seeds. In spring and early summer they feed on bulbs and young shoots of trees. At this time of year they will also take birds' eggs and nestlings. They can have a devastating effect on songbird populations and they compete with hole-nesting birds for breeding sites. They have also, to an alarming extent, replaced native red squirrel populations throughout the British Isles. Furthermore, they damage horticultural crops, garden produce and amenity trees. Nevertheless, they are now part of our wildlife and many people enjoy their presence.

In Britain the grey squirrel has no natural predators nor is it seriously affected by disease. Population levels are only controlled naturally by the availability of food during the winter months. Good seed years lead to high winter survival rates and in many locations winter feeding of pheasants has the same effect. Artificial feeding of squirrels in urban parks and gardens also helps support high population levels.

Grey squirrels frequently cause significant damage to trees by bark stripping. This occurs mainly between the end of April and the end of July. It can occur anywhere on main stems from base to crown. Few trees die but many produce degraded timber through rot and broken tops. This can be a source of danger where there is public access. Sycamore, beech, oak and sweet chestnut trees of five to 40 years of age are the most vulnerable.

The wholesale elimination of grey squirrels is neither practicable nor desirable (except perhaps in some areas to conserve red squirrel populations). Damage control can be limited to woods where timber production is an objective. In small isolated woods control may not be necessary because small areas will not hold large populations of squirrels. Control must be carefully targeted and selective. It is best carried out in and around vulnerable crops from April to July. Control should be carried out by cost-effective, humane and selective methods. Shooting and drey poking, which can only be carried out effectively in winter and early spring before leaves flush, is unlikely to control numbers sufficiently to reduce damage. Surviving squirrels have a better food supply and may produce larger and stronger litters. This, combined with colonization into vacant territories from uncontrolled areas, can lead, by early summer, to population levels as large as those that would have existed had no shooting been undertaken. Shooting used in conjunction with other methods of control, however, can be useful.

Cage trapping (single or multicatch) is effective, though it is a labour-intensive method since the traps, by law, must be inspected at least once every day. Multicapture traps are usually set out at a maximum density of one per hectare. For small woods and where labour costs are not a consideration, this is probably the best method. Spring trapping using humane traps in tunnels (as required by law) is also effective. They too must be visited daily.

Research has shown that the use of warfarin-treated bait in hoppers is the most cost-effective method of control.[15] The 1973 Grey Squirrel (Warfarin) Order permits, with the exception of a few specified counties, the use of warfarin in hoppers of approved design (that cannot be accessed by red squirrels). Hoppers should be set out approximately 100 to 200 metres apart (three to five per hectare). Each hopper should be positioned at the base of a tree, where it should be firmly secured. If badgers are present, hoppers should be sited a metre or so above ground, either in a fork of a tree or on a stand. Special care should be taken where the (legally protected) dormouse is found, and warfarin should never be used where pine martens (an endangered species) are present.

A national proficiency test certificate should be obtained by anyone using warfarin. The Forestry Authority encourages the setting up of grey squirrel management groups where there is a strategic need to protect large areas of vulnerable woodland. These groups can offer information and advice, and some may be able to advise on service contracts with qualified forestry practitioners. To ascertain whether a group operates in any particular region, the nearest Forestry Authority office should be contacted.

Rabbits

Wild rabbits have once again become a major pest in woodlands.[16] Serious damage to young plants and pole-stage trees of many species occurs throughout Britain. Rabbits eat young tree seedlings, cut off shoots and browse on branches. They help prevent woods from regenerating naturally and they will even damage the thick bark at the base of semimature trees, resulting in serious timber damage or, if ring-barked, death. Browsing damage by rabbits can be distinguished from that caused by deer and sheep: deer have no incisors and leave a ragged cut, whereas rabbits leave a clean and usually oblique cut that is rarely torn.

Earlier this century rabbit clearance societies were common. Countless numbers of rabbits were shot, ferreted and trapped – for food, sport and to protect crops. But changing tastes, greater affluence and the population drift from country to town have played in the rabbit's favour. Even myxomatosis gave only a temporary respite; this disease is no longer a major controlling factor.

There is a statutory requirement for all occupiers of land, whatever its use, to control rabbits. Because of the scale of the effort required and the rabbit's inherent capacity for population increase, complete eradication is impractical. Rather, the aim should be to reduce and maintain rabbit numbers to levels at which damage is economically acceptable. The most effective time for control is from November to February.

The control of rabbits in woodlands can be an uncertain, expensive and time-consuming business. Methods of control include fencing, tree guards and the use of repellents. The Ground Game Act of 1880 gives every occupier of land a limited right to kill and take rabbits and hares, concurrently with the right of any other person entitled to do so on the same land. An occupier may use any legal method to kill rabbits such as gassing, trapping, ferreting, shooting, snaring, netting and, with the exception of shooting, he may authorize other persons to assist him. The act exempts an occupier, and persons authorized by him to kill rabbits, from the need to hold a game licence.

More recently, a simple and effective system for eliminating rabbits has been developed using fumigation pellets. These pellets react with moisture in the soil and in the air, releasing a fatal toxic gas in rabbit burrows. The pellets are set in place by means of various specially designed applicators. Fumigants must be approved under the Control of Pesticides Regulations 1986 and must be used according to label instructions. Gassing is a two-man job and must not be carried out in wet or windy conditions. Fumigants can be lethal to humans. The method should be applied only by an operator trained in its use and familiar with the appropriate cautionary measures. One-day training sessions on rabbit control are available from registered instructors, and certificates of competence are awarded in appropriate cases to those completing the course.

Hares

Hares are far less numerous than rabbits. They use woodlands mainly for shelter – their preferred habitat being agricultural land. They damage young trees in much the same way as rabbits. In winter, they often systematically work their way along a row of trees, leaving the bitten off shoots lying on the ground. Damage is seldom widespread, though it can be severe on a local scale. Although they can be successfully snared, the usual method of control is by driving and shooting. The cooperation of neighbours in forming shooting lines will add greatly to the effectiveness of the operation.

Deer

There are two species of deer native to Britain – red and roe – and four introduced species – fallow, muntjac, sika and Chinese water deer. The wolf, which helped to control deer numbers naturally, has been exterminated, and changes in landuse practice, especially an expansion in woodland cover, have increased the availability of deer habitat, stimulating population increase and spread.[17,18] There are probably more deer in the countryside today than at any time in the past.

Too many deer cause damage to trees and to wildlife habitats in woodlands.[19] They do this by browsing leaves and shoots (which can prevent trees from regenerating), by stripping bark for food (which can kill trees) and by fraying tree stems (which they do to mark territories and to clean antlers by rubbing off the velvet). Damage is controlled by fencing, dead hedges, chemical repellents and population control. Advice should be taken in good time; control should start 18 months to two years before any restocking work is planned and should form part of the woodland management strategy. In some areas where browsing pressure is intense, owners and managers can cooperate to organize deer management using the services of qualified wildlife rangers.

AFTERCARE

Beating Up

Beating up is the replacement of plant losses. An assessment of missing trees is generally made at the end of the first growing season. Beating up is only worth doing in the first two or three years after planting. Left later than this, the replacements will have little chance of catching up with the original trees. Where trees are spaced three metres apart or wider, all missing trees should be replaced. Elsewhere it is usually sufficient to substitute one replacement for every two failures.

Some losses after planting are almost inevitable but with relatively close spacings (up to about two metres), scattered losses of up to 20 per cent may be tolerated and no replacement planting need be carried out. At spacings greater than this, individual losses will be more apparent.

Weeding

Weeds reduce the survival and growth of newly planted trees by competing for light, soil moisture and nutrients.[20] Early and effective control is the key to successful tree establishment. Given effective preplanting treatment, weeding should only be necessary in the second and third years after planting.

The most cost-effective method of weed control is by use of herbicides. A weed-free area of about one-metre diameter around each tree should give significant results. Control over the whole area is generally unnecessary and environmentally undesirable. Woody weeds (coppice regrowth, natural regeneration and root suckers) will need to be dealt with wherever they are not wanted.

Treeshelters will protect trees from herbicides. A medium volume application using a knapsack sprayer can be applied with care to weeds outside the shelter. The time of weeding is important. Done too early, another weeding may be required before the season's end. Residual herbicides (those that remain active in the soil for some time after application) should be applied in winter. For other herbicides, late spring or early summer is best. Always seek specialist advice before using chemicals.

Herbicide Application

No specific advice on herbicides and their application is given in this book. The large number of weed and crop species, the variety of suitable herbicides, the application methods available and the various statutory controls make the subject a complicated one. The reader is referred instead to *Forestry Commission Field Book No 8 – The Use of Herbicides in the Forest*. Members of the Timber Growers Association have access to a pesticides helpline that can provide up-to-date information and advice on various aspects of pesticide and herbicide application.

Cutting weeds using hand tools or machinery will reduce competition for light but will do little to reduce competition for moisture and nutrients. It may even increase it, since regular cutting stimulates the growth of weeds and thus increases water uptake. On the other hand, relatively little moisture evaporates from bare soil once a layer of dry soil forms on the surface.

Bracken areas can be dealt with by 'bracken whipping'. This involves whipping off the heads of the tender young fronds early in the growing season before they have hardened off.[21] Whips can be easily made from slender, flexible, woody stems cut on site. Bracken whipped in June will have regrown by August and, although less vigorous, may need further control in some areas. Whipping is a suitable method for volunteers as it does not involve the use of sharp tools. However, operators should wear face masks and plastic face shields to comply with current recommendations.

Cleaning

Cleaning is the removal of any unwanted woody growth from around well-established trees. It can be a costly operation. It is not always necessary, though where unwanted growth is prolific, neglect can lead to poorly developed trees. Cleaning can be undertaken using conventional hand tools, chainsaws or a clearing saw (a small-diameter circular saw incorporating a long extending handle and body harness).

Areas of natural regeneration may have to be cleaned to remove any unwanted species and to reduce any overstocking. Unwanted coppice regrowth is not easily dealt with. Ideally, the freshly felled stumps should be sprayed with an appropriate herbicide prior to planting. Cutting with hand tools is rarely effective. The need for cleaning will diminish as the trees begin to suppress the undergrowth.

Dealing with Sycamore

Sycamore seed dispersal and germination is efficient and generally successful. It is a vigorous colonizer of both open ground and shaded woodland, and although it produces a very useful timber, its presence in great numbers does not always find favour with conservationists.

Sycamore can be difficult to control. It is only practically eliminated on sites where it is sparsely distributed and, even then, treatment will need to be intensive and may prove costly.[22] Control of the main seed-producing trees is essential. Trees can be killed by ring-barking provided it is thoroughly executed; ineffective attempts can lead to prolific seed development. Trees must be felled as close to the ground as possible and the freshly cut stumps should be treated with an appropriate herbicide to which a dye has been added to help identify treated stumps. Often, some coppice regrowth will occur. This can either be recut and the stump treated again or the fresh summer foliage can be treated with a suitable herbicide.[23]

Dealing with Climbing Plants

Climbing plants (such as honeysuckle, ivy and old-man's beard) are a natural component of British woodlands. They provide nesting sites and shelter for birds and they attract insects, providing birds with food.

Climbers do not kill trees, they use them for support, and most healthy trees remain relatively unaffected by their presence.[24] It is questionable whether climber roots compete to any significant extent with tree roots for moisture and nutrients. They may possibly weaken trees by reducing their capacity to photosynthesize, but in an old experiment, ivy was cut on half the trees in a crop and left on the rest. After 40 years' growth, ivy covered the trunks and main branches; however, when the trees were felled at 52 years, no difference could be found in height, average girth or volume.[25] Climbers can certainly decrease wind resistance, leading to damaged branches or even whole trees blowing over. Their presence can also hinder felling operations, adding to harvesting costs.

The control of climbers can be confined to timber-producing areas or where they occur on rarer species or specimen trees. They can be preserved, however, along woodland edges and rides for the considerable environmental benefits provided. If climbers must be cut, this should be done low down and care must be exercised not to damage the tree. There is usually no need to unwind or remove them.

Pruning

Pruning involves the removal of tree branches. It facilitates access and increases potential profitability by minimizing knots in the most valuable part of the tree – the lower stem. Pruning takes several forms.

Formative Pruning

Pruning for the first 15 to 20 years of a tree's life can transform the quality and value of the mature tree. If side branches are not removed, they grow or they die back. If they die back they result in dead knots that lead to defects in the finished timber. Pruning needs to be carried out when branches are small enough to heal quickly.

When pruning, about one half to two-thirds of the tree should be left with green branches so as not to seriously reduce vigour. Broadleaves can be pruned when the stems are six to eight centimetres in diameter at breast height. It is important not to cut branches flush to the stem because this removes many of the tree's natural defences against decay and infection (and as such may reduce timber value).[26]

A pair of heavy-duty secateurs is best for pruning young

Figure 3.9 *Long-Handled Pruning Saw*

broadleaves. For older broadleaves and conifers, a pruning saw with a sickle-shaped blade is the usual tool. Chainsaws are not recommended; even in skilled hands it is difficult to avoid damaging the bark of the trees. Large branches are best undercut first to reduce the chances of the wood tearing. The optimum time to prune broadleaves is the subject of current research. Wild cherry is best pruned between June and August to minimize risk of infection from bacterial canker and silver leaf disease. Sycamore is best pruned in February to avoid blue stain.

Brashing

The pruning of conifers up to a height of about two metres is termed brashing. This is done to improve timber quality on the most valuable part of the stem, to facilitate access for inspection, to improve sporting potential and to help prevent ground fires from becoming crown fires. Trees are ready for brashing when the stem averages about 15 centimetres dbh. The work is customarily undertaken in winter; summer sap flow can lead to unpleasant working conditions and leaves sticky and unsightly exudations on the bark.

Brashing and pruning are labour-intensive operations that can be confined to selected trees or occasional rows only. The operation is often combined with cleaning or first thinning. Brashing and pruning can lead to reduced costs of subsequent operations such as marking trees, thinning, felling and trimming out, but as these works generally take place early in the life of the crop, due increases in compound interest must be allowed for.

High Pruning

High pruning is the progressive debranching to heights above two metres to produce knot-free timber of greater market potential. The operation is usually confined to selected trees. Pruning to around four metres can be carried out using a conventional pruning saw, or a chisel pruner, fitted with a long handle. For branches above this a ladder and short-handled pruning saw can

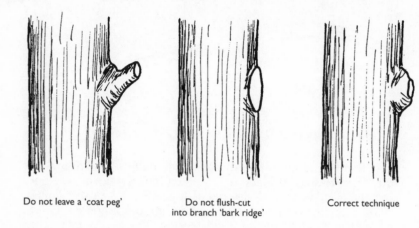

Do not leave a 'coat peg'   Do not flush-cut   Correct technique
into branch 'bark ridge'

Figure 3.10 *Pruning Technique*

be used. Alternatively, or if a large number of trees have to be pruned, it will probably be worth investing in one of a range of specialist pruning saws. The longest – a telescopic saw – has an adjustable head and an amazing 6.5-metre reach (about 8.5 metres allowing for the reach of the operative).

High pruning is ideally done between September and December. It is important not to prune a tree when it is flushing. Between a half and two-thirds of the tree should be left with green branches. Whether high pruning is a cost-effective operation has long been debated. It is not usually profitable if someone has to be paid to do the work, but where it can be done with no cash outlay (for example, by the owner) it can be worthwhile. There is a greater price differential between clean and knotty broadleaved timber than there is in the case of conifers.

SOME IMPORTANT WOODLAND DISEASES

Honey Fungus

Honey fungus (*Armillaria* species) is a root pathogen very common in woodlands, parks and gardens.[27] It derives its name from the honey-coloured toadstools that sometimes appear in large clusters around the bases of affected trees in autumn. The toadstools, which have a whitish frill just beneath the cap, perish with the onset of the first frosts. In the absence of toadstools the disease may be recognized by the appearance of black bootlace-like strands (rhizomorphs) that spread through the soil and under the

Table 3.6 *Susceptibility of Trees to Honey Fungus*

Particularly Susceptible Species	Species Showing Some Resistance
Birch	Yew
Walnut	Oak
Willows	Lime
Spruces	Hornbeam
Western red cedar	Beech
Western hemlock	Ash
	Blackthorn
	Box
	Holly
	Hawthorn
	Larch
	Douglas fir

bark of affected trees, sometimes in a tough, flat interlaced mat. Furthermore, a characteristic creamy white, soft fungal mass of mycelium may sometimes be found beneath the bark of decaying roots and lower stems.

In Britain the disease rarely causes extensive damage. Rather, the typical effect is to kill trees in ones and twos within a small area. Death can appear to be sudden, or it may follow a gradual die-back. A wide range of trees and shrubs can be affected.

The practical difficulties of controlling honey fungus are considerable. The stumps and roots of all affected material would have to be removed and destroyed, and to be on the safe side the soil would need to be replaced before replanting with resistant species. Such measures are unlikely to be practical. Honey fungus hot-spots might be worth considering as potential nature conservation sites.

Dutch Elm Disease

Dutch elm disease (*Ceratocystis ulmi*) is something of a misnomer since it is neither restricted to Dutch elms nor did it originate in Holland. The disease was discovered in France, identified in the Netherlands and probably originated in Asia.

Dutch elm disease is one of the most serious tree diseases. It is caused by a fungus and is spread by various elm bark beetles (mainly in the genus *Scolytus*). In Britain the disease was first noted in 1927.[28] It caused widespread deaths of elms in southern England but many partly affected trees subsequently recovered. In the 1960s a more virulent form of Dutch elm disease struck the

British elm population. This time the effect on the landscape of southern Britain was devastating, with over 20 million elms killed by 1980. The disease is now present in Scotland. The elm population in Britain is very diverse but all subspecies and varieties are susceptible to the disease to a greater or lesser extent. The native wych elm is not immune although it does exhibit better survival rates in some areas.

The first sign of the disease is often a yellowing or browning of the foliage on part of the tree. Affected branches begin to die back from the tip. In a severe attack the entire tree is killed before the end of the summer, but even if it survives it may well die in the following spring. During summer and early autumn, adult female beetles bore into the bark of moribund elms, tunnel out a gallery and lay their eggs along its length. The resultant larvae make secondary tunnels at right angles to the main gallery. They then pupate and eventually emerge as young adults from small circular holes in the bark in the following spring. These young beetles can carry spores of the fungus on their bodies. Before breeding, they often fly into the crowns of healthy elms to feed on the bark in the forks between twigs. In the process, the damaged tissue may become infected with spores. The disease can also pass from diseased to healthy tree by root contact. Following death, large elms can remain suitable for beetle breeding for up to two years, though the thinner bark of smaller trees is likely to become unsuitable after just a year.

Since the 1970s there has been much regeneration of elm from root suckers in areas devastated by the disease. These new trees are genetically identical to the parent trees and, therefore, are no more resistant to the disease. Many have attained a height of ten metres or more but since 1990 the disease has returned to kill much of this regrowth. There will probably continue to be a cycle of regeneration followed by outbreaks of the disease. The cycle may be as short as 12 years in southern areas.

A number of control measures have been attempted, including sanitation felling (the prompt felling of diseased trees and the rapid and safe disposal of potential beetle-breeding material) as well as the use of insecticides and the use of fungicides. In 1994, east Sussex, largely due to an enthusiastic and sustained control campaign, had about the only significant population of English elm remaining.[29] In a few rare instances small groups and individual trees show no signs of being affected and work is being carried out on the selection and breeding of elms for resistance to the disease. A recent initiative in Britain is an attempt to introduce disease resistance by genetic engineering.[30]

Owners opting to retain elm in their woods must expect

failures. Young growth is best respaced to at least nine to 12 metres apart to leave sufficient room for interplanting with other species. Otherwise, suckers could be coppiced on an eight- to ten-year cycle on the chance of keeping the disease at bay. All cut elm material must be immediately debarked or destroyed by burning or chipping. No attempt should be made to store the material for firewood as it may already be infested with bark beetles. Timing the operation to coincide with bonfire night is not a bad plan.

Fomes Rot

Fomes (*Heterobasidion annosum*) is a major fungal disease of commercial conifers. It is also an occasional problem of broadleaved and conifer amenity trees, causing rot in the roots and lower stems. The disease is present in most areas that have previously carried conifer crops. It spreads by root contact from infected to healthy trees and new infection arises when airborne spores alight on the freshly cut stumps of felled trees. Treatment of conifer stumps, to protect against the spread of the disease, is common practice in British forests. Control is effected by coating the stumps with a solution of urea immediately after felling. On strictly commercial grounds it may only be justified on sites with a high risk of the disease, though on other sites non-market benefits may be sufficiently important to make treatment worthwhile.[31]

Pine stumps (but not those of other conifers) can be treated with commercially produced spore suspensions of another fungus that prevents the entry of fomes and decays the stump without posing any threat to surrounding trees.

PLANTING NEW WOODS

This book is about the management of existing woodlands rather than the creation of new wooded areas. Nevertheless, much of what has been written concerning the replanting of felled sites applies to the planting of new land. The main areas of concern in which planting and replanting are likely to differ are discussed below.

Exposure

Most new planting will be on former agricultural land which, even in lowland Britain, can be very exposed. Persistent winds lead to deformed trees and generally poor growth. Replanting sites in woodlands are likely to be comparatively well sheltered.

Weeds

Former arable land will often hold a bank of weed seeds that will germinate and grow rapidly when routine weed control ceases. A comparable situation arises with permanent grassland, which will rapidly grow tall and rank when livestock is excluded. Felled sites in woodlands are usually much less weedy.

Rootability

Trees need an ample depth of soil if they are to grow satisfactorily. Repeated cropping of arable land sometimes leaves a compacted zone of soil at the base of the ploughed layer (a ploughpan) that can hamper tree root development and may eventually lead to crop instability. The answer is to break up any pan using a tractor-mounted deep-tine cultivator. Woodland soils do not present this problem.

Obstacles

Obstacles on the planting site are not only hazardous, they can reduce operator efficiency, leading to higher unit costs. Other than for small stones, former agricultural land is usually free of impediments, whereas felled sites in woodlands will certainly contain tree stumps from the previous crop and may also be strewn with branches and tree tops.

Watertable

When trees are felled there is a general rise in the watertable. It may be necessary to dig cut-off or surface water drains, and trees may need to be planted on mounds to raise roots above the watertable. Former agricultural land will rarely have these problems.

Site Disturbance

Harvesting operations in woodlands can leave deep ruts which, particularly on level sites and on clay soils, can render the site compressed and waterlogged. Soils left in a compacted condition can be a cause of heavy plant losses. New planting sites will not pose these difficulties.

Frost Pockets

The felling of groups of trees within a wood may allow the formation of localized frost pockets. Badly drained sites can exacerbate the problem.

Grant Aid

Grants for new planting are generally higher than those offered for replanting.

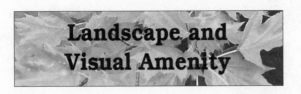

4

Landscape and Visual Amenity

Woodland owners are enjoying their woods now more than ever before. As well as income from timber production, there is a growing recognition and appreciation of the many other benefits that small woods have to offer, such as the maintenance and enhancement of landscape and visual amenity. Landscape can be described as an extensive area of scenery, viewed from any location; visual amenity is a pleasant prospect seen at close quarters.

Some felling may be unavoidable if a wood is to be improved, but the scars left after harvesting can be visible for miles. In a large woodland, robust flanks of trees can be retained to conceal the disturbance, but in a small wood there will usually be no room for such manoeuvres. Narrow belts of trees are useless as an aid to screening and are vulnerable to damage by wind. While operational work (such as thinning and felling) can occasionally provide opportunities to improve any unsympathetic layout, careful planning and appropriate management are clearly essential if the benefits that small woodlands bring to the landscape are to be fully developed.

ASSESSING LANDSCAPE CHARACTER

In considering how well existing features fit into the countryside, some form of visual assessment is necessary. But just what does constitute an attractive landscape? How is high visual amenity to be judged? As concepts they are almost impossible to interpret since they are to do with one's own individual instincts, experience, culture and education. A sight regarded as enchanting by one observer can set another's teeth on edge. The artist Lowry and his devotees were captivated by grimy, industrial landscapes; some committee or other thought it appropriate to plant trees in the form of an enormous 'EIIR' on a prominent Welsh hillside; concrete cows welcome visitors to Milton Keynes; and a huge topiary ox is planned for the approach to Oxford city. Such cases do not win universal acclaim, but it is not possible to please everyone when so much is a matter of personal taste.

Some qualities, however, seem to be appreciated by most people. The sight of natural scenery – mountains, rivers, hills, valleys, trees and woods – gives immense pleasure, fills us with wonder and invariably brings a sense of peace and well-being. In creating and managing woods, the skill must be to establish and maintain scenery that looks as if it came about by natural means.

MANAGING CHANGE

People generally dislike change and react unfavourably to any alteration in familiar surroundings. But change is inevitable in the working countryside and woodlands are part of the scenery. Woodlands have been managed for thousands of years to meet the needs of society. They were, and are, used to produce timber and other wood products, to afford shelter and to facilitate sporting activities. But, working woodlands can still be attractive. The skill is to emulate natural conditions as far as possible. The Forestry Authority publication *Lowland Landscape Design Guidelines* offers some useful advice.[1]

One approach is to identify the main viewpoints from which a particular woodland is most frequently seen (for example, a dwelling, picnic site or lay-by). From these focal points sketches can be made, so that the visual impact of impending work can be foreseen and any necessary design modifications considered in advance. While individual circumstances will obviously vary, the following guidelines may be found useful.

- Provided crops are stable, take advantage of thinning operations to correct poor landscape design (for example, by modifying woodland edges and realigning rides).
- In mixed woods, influence seasonal changes of colour by adjusting the proportions of species when thinning.

TRAINING HELPLINE

For the answer to all your questions about training issues in agriculture and horticulture

0345 078007
... at the cost of a local call

- Where line mixtures have been planted, soften abrupt changes of colour and form by breaking up rows of alternating species.
- Avoid the systematic removal of complete rows when thinning in areas of high landscape sensitivity.
- Along woodland boundaries, rides, glades and power-line corridors, develop a cascade-effect transition from low vegetation, through shrubby growth and short trees, to the main crop trees in the background.
- Vary the width of woodland rides and power-line corridors by cutting an irregular distribution of sweeping bays and 'scallops' of different sizes.
- Create informal shaped glades where roads or ride systems converge.
- Design felling coupes in scale with the landscape. Very small coupes may be appropriate in some instances but they can create a 'currant bun' appearance on hillsides.
- Design coupes so that they taper away from main viewpoints.
- Design coupes with interesting, irregular outlines.
- Vary size and shape where more than one coupe is planned.
- An irregular distribution of coupes will help create diversity.
- The full extent of felling on hillsides can be disguised by cutting elliptical coupes with the long axis running along the contour.
- Plan felling coupes to take advantage of interesting views.
- Any trees left standing in felling coupes should be in robust groups or drifts. Individual or widely spaced trees are prone to wind damage and desiccation.
- Avoid retaining a scatter of standing trees on the skyline. They give an unpleasant saw-toothed effect.
- In very small woods being managed for timber production, clearfelling the entire area and replanting may be the only visually acceptable alternative.

- To give a variety of heights, plant trees of different growth rates.
- Choose treeshelter colours to blend in with the surroundings.
- Avoid planting in straight rows wherever possible.
- Where straight rows are essential, align them at right angles to main viewpoints and stagger the distance between trees.
- Consider additional planting to increase size and scale and to improve the shape of woods.
- Diversify a monotonous woodland edge by planting a few irregular shaped groups as extensions to the wood.

5

Wildlife Conservation

Wildlife conservation will be the main management objective for many woodland owners. Occasionally, the aim will be to encourage some particular species, but more often it will be to enhance wildlife interest in general. This is best achieved by increasing biodiversity.

Woodland species have adapted to the natural processes that took place in the wildwood before man cleared and modified it. Some woodland management is now necessary due to the loss of many natural processes and because woods are now generally small and fragmented. This fragmentation means that some species have great difficulty in dispersing to new suitable areas. Management is needed to artificially maintain suitable conditions for a wide range of species and to control others that can damage woodlands.

THE WOODLAND ENVIRONMENT

Though large woods are generally of greater wildlife value than small woods, a small diverse wood can contain more species than a large uniform one. Even so, the typical small wood cannot possibly sustain all the woodland species worthy of conservation, so it is essential to identify the important features before any work starts. Only when a thorough assessment of the wood has been made will it be possible to consider the most appropriate form of wildlife management; however, whatever work is eventually planned, the one overriding principle should be to avoid drastic change wherever possible. In broadleaved woods three options are generally available:[1]

- maintenance or reinstatement of traditional management practices (for example, coppicing);
- introduction of new management regimes (for example, group felling and restocking);
- non- or minimum intervention.

The choice will turn on such factors as the:

- owner's objectives;
- owner's financial position;
- historical management of the area;
- overall conservation value of the wood;
- presence of rare or interesting species;
- size of the wood;
- potential markets for timber or woodland produce;
- existence of constraints (such as heavy deer pressure).

Generally, the better the conservation, the worse the finances. On a wildlife reserve, for example, deadwood is more valuable than veneer-quality timber. Nevertheless, modest conservation measures can often be incorporated into a management plan at little extra expense. Timing the conservation work to coincide with woodland operations is the key to minimizing costs. Comparatively minor adjustments to practices can be of considerable value for wildlife conservation. But care must be exercised – direct management to benefit one species may be the very opposite of what another, perhaps equally important, species requires to survive.

Dying and deadwood provides one of the most important resources for animal species in woodland.[2,3] Something in the order of one third of all woodland birds nest in holes in old trees. Oak, ash and beech only begin to provide rot-holes when they are around 100 years old. Dying and deadwood is extremely important in terms of the invertebrates that it supports. It also provides a niche for a range of harmless fungi. The majority of fungi species are to be found in woodlands.

Britain is said to have the largest and best examples of extremely old trees in north-west Europe.[4] These veteran trees are capable of supporting an immense range of wildlife, from such large and obvious creatures as owls, woodpeckers and bats, to minute forms of life and elaborate communities of lichens, mosses and fungi that play a vital role in deadwood decomposition. With one or two exceptions (for example, dead elm which can harbour the beetles that spread Dutch elm disease) there is little or no economic or plant health risk in retaining deadwood.

Below are some guidelines for maximizing wildlife habitats in small woods. Not all the advice will be appropriate in any one place.

WOODLAND COMPOSITION AND STRUCTURE

- Aim for a diverse woodland structure with well-defined ground, field, shrub and tree layers. This will ensure a wide range of light conditions.
- Where they threaten to become dominant, control sycamore, rhododendron, snowberry, Japanese knotweed and other invasive species. Left to their own devices they can shade out less vigorous native trees and shrubs.
- Control grazing levels. Some woodland plants benefit from a light grazing regime but prolonged periods of heavy grazing can prevent tree regeneration.
- Reintroduce coppicing where there is evidence of past use. Adequate deer management is essential. If coppicing is not commercially viable then the owner must have the resources to maintain it. Rideside coppicing is more practical where resources are limited.
- Reintroduce coppice-with-standards where there is evidence of previous use. It provides, in a relatively small area, a great variety of valuable wildlife habitats.
- Consider creating new coppices where none exist at present.
- Coppice elm regrowth. This will keep branches young and may help keep Dutch elm disease at bay.
- Utilize natural regeneration in appropriate places rather than replanting. Encourage regeneration (where adjacent land is in the same ownership) to increase the area of woodland.
- Preserve any old walls and abandoned buildings in the wood. They provide ideal habitats for moss, liverwort and lichen communities.

HEDGES AND EDGES

- Create wildlife corridors to attract wildlife into and out of woodland. Woods with a network of interconnecting hedgerows and planted belts have a greater potential for attracting and holding wildlife than woods isolated as islands within a sea of grass or arable land.

Deadwood is a natural component of woodlands. A proportion should be left for conservation benefits.

- Relay woodland hedges regularly. Properly maintained hedges are of greater wildlife value than neglected ones. They provide nesting sites for birds, a rich supply of insects on which birds and bats can feed and a warm sheltered environment within the wood. (Grants may be available for hedgerow restoration work. Contact local FWAG officer.)
- Allow occasional hedgerow trees to grow to maturity. The presence of trees within hedges will add to the range of birds present.
- Cultivate a cascade edge effect on rides, wood edges and boundaries. A graded edge will provide good shelter, a wide range of habitats and a more pleasing aspect.
- Encourage nectar-producing trees such as hawthorn, sallow and blackthorn on boundaries and ride sides. These attract insects.
- Allow the development of a few blackthorn thickets. They are particularly valuable as songbird and butterfly habitats.
- Consider repollarding old pollards (see Chapter 2).
- Create occasional new pollards to provide a succession.

RIDES, ROADS AND OPEN SPACES

■ Widen woodland rides and paths. This is one of the surest ways of enhancing the wildlife interest of a wood. It will also improve woodland amenity and provide better facilities for timber-harvesting operations.

■ Develop a number of wildlife glades. Areas in full sunlight become rich in wildflowers and grasses that will support a wide variety of butterflies and small mammals.

■ Where rides intersect, create a large open space by cutting back the corners.

■ Link rides, glades and ponds to provide conservation corridors within woods.

■ Mow rides at irregular intervals. Mowing alternate sides every other year, preferably late in the season, will allow flowering plants to develop. Leave an outside strip to be swiped every third year to provide diversity. Remove the cuttings. If left, the nitrogen created by decomposition will encourage grass to grow, which can smother butterfly-friendly plants.

■ Avoid mowing under trees. Mowing under trees can be counterproductive. It removes cover, increases surface vegetation transpiration rates, thus depriving trees of moisture, and often results in bark damage to trees.

■ Ensure rides are at least as wide as the tallest trees on the rideside. Progressively cut back overhanging trees as they grow taller.

■ Redesign straight rides to incorporate sweeping curves and shallow bays to give a scalloped effect. This provides more sheltered conditions.

■ Use only local stone for surfacing roads. Introduced material may change local conditions and alter plant communities.

■ Utilize and develop land under power lines and other wayleaves as conservation areas.

■ Preserve any rock outcrops. They provide additional habitat for wildlife.

■ Do not fill in old quarries or hollows by dumping. Preserve as much variety in the wood as possible.

■ Herbicides used early in the year to control bramble can encourage herbaceous plants.

ANCIENT TREES AND DEADWOOD

■ Resist any urge to tidy up woods. The presence of dead trees and large branches is often seen as a sign of poor management. In woods devoted to wildlife conservation, nothing could be further from the truth.

- Retain small groups of low-quality trees to be left to grow old and die naturally.
- Where few dead trees exist, consider ring-barking any unwanted trees to provide deadwood.
- Where there is a shortage of fallen deadwood, bring in small heaps of unwanted timber. Leave shattered ends to the logs, rather than square sawn, to encourage rot. Site the heaps in sheltered and shady places to create better conditions for decay. Deadwood in open places tends to become dry and hard with less abundant invertebrate life.
- Do not attempt to repollard a very old tree without first consulting a specialist.

MANAGEMENT OPERATIONS

- When trees are pruned, heap the branches in windrows. This will allow the ground vegetation to develop between the rows and will provide low-level shelter, cover and nest sites. Windrows can, however, harbour rabbits.
- Consider high-pruning final crop trees. This lets in more light and facilitates movement for birds such as hawks and owls.
- Where major operations are planned, identify some sections of the wood to be left undisturbed, to serve as refuges into which wildlife can retreat while work proceeds.
- Where possible, avoid felling, thinning and hedge laying in the main bird nesting season from March to July.
- Where possible, restrict operational work to no more than a third of the total woodland area in any one year.
- Thin rideside edge trees more heavily. Although traditionally edge trees were thinned less heavily for better shelter, this lets in more light, allowing the ground vegetation to flourish. Care must be exercised, however – a sudden change of practice can result in wind damage to crops.
- When thinning broadleaved mixtures, favour native species for retention wherever possible. Favour oak in particular: it is outstanding for the wildlife it supports.
- When thinning conifer and broadleaved mixtures, favour the broadleaves provided this is not incompatible with the objectives – but retain a few conifers to add diversity and provide warm winter roosting.
- Select for long-term retention any less common species. Examples might include crab apple, spindle, guelder rose,

wild privet, wayfaring tree, buckthorn and wild service.

■ Use as few timber extraction routes as possible. Use mats of brash to reduce soil damage; use temporary culverts where timber has to be extracted across soft-banked streams.

■ Consider the use of horses, rather than tractors, for timber extraction.

■ Do not intentionally or recklessly damage or obstruct a badger sett which shows signs of being used. It is a criminal offence to do so. Create protection zones of 20 metres around the setts and avoid the breeding season when carrying out woodland operations.[5]

■ Install badger gates where their runs cross proposed fence lines.[6]

■ Consider leaving a proportion of the wood permanently unmanaged. Areas of permanent woodland cover encourage the growth of mosses and liverworts. Plan non-intervention areas where these species already occur.

■ Try growing your own trees and shrubs from locally collected seed of species that are native to the area. Many species germinate freely and the activity will give hours of pleasure.

HABITAT CREATION

■ Leave heaps of decaying vegetation or sawdust in a sunny site for grass snake and slow-worm breeding sites. They lay eggs in June and July for hatching in August and September.

■ Rough unmanaged fields adjacent to woods make good adder territory.

■ Encourage reptiles by laying corrugated sheets in sunny glades.

■ Rotational strips of conifers, such as might be created for Christmas tree production, can also provide good reptile habitats.

■ Encourage dormice, if present, by the provision of thick cover such as coppice, scrub or honeysuckle. Dormice are harmless and are a protected species in the UK. They are uncommon. They need ample supplies of seed- and fruit-bearing trees and shrubs, particularly hazel, to provide autumn food. Nest boxes can be provided.

■ Encourage harvest mice by leaving patches of thick vegetation and by spot weeding around trees rather than complete weeding. They are found along ridesides and in new plantations. They cause no significant damage to woodlands.

■ Where there is a shortage of natural sites, provide artificial holts for otters. Otters occur in all types of woodland, provided there is freshwater in the form of streams, rivers or large ponds.

■ Erect specially constructed summer roost and winter hibernation boxes high up on trees for bats.[7] Many of our 15 resident bat species are dependent on woodlands for foraging and roosting. Ponds are especially important as foraging habitats.

■ Encourage birds by providing nest boxes for various species. The Royal Society for the Protection of Birds (RSPB) will provide advice on siting and design.

■ Accept the presence of climbing plants. Ivy-covered trees are particularly favoured by long-eared owls for roosting and honeysuckle is a favourite lair of dormice.

■ Increase butterfly populations by encouraging associated plant species, particularly alongside wood edges, rides and glades. Table 5.1 lists a number of butterfly species that exhibit a strong association with certain trees and plants.

WETLAND HABITATS

■ Conserve any damp hollows or marshy areas. Many wetland animals need dry land during part of their life, so wet areas within woods are more important than most habitats in this respect. Woodlands also provide the sheltered environment preferred by many of our 38 dragonfly and damselfly species.

Table 5.1 *Butterfly and Plant Associations*

Butterfly	Associated Species
Brimstone	Buckthorn and alder buckthorn
Brown and black hairstreak	Blackthorn
Comma	Elm, nettles, hops
Duke of Burgundy	Cowslip and primroses
Fritillary butterflies	Violets
Holly blue	Holly, ivy and others
Purple hairstreak	Oak
Purple emperor	Sallow
White admiral	Honeysuckle
White letter hairstreak	Elm
Wood white	Meadow vetchling, vetches and bird's foot trefoil

- Where a stream runs through a wood, aim for intermittent light and shaded conditions and encourage only light-foliaged broadleaved trees that are native to the area. Birch, willow, rowan, ash, hazel and aspen are often appropriate depending on the locality.
- Most of our native amphibians prefer wet areas surrounded by rough grass and scattered trees. Leave piles of fallen or cut logs near ponds for newts to hide and feed in.
- Avoid feeding fish or wildfowl where algae is perceived to be a nuisance. It exacerbates the problem.

POND MANAGEMENT

The number of ponds in Britain has declined by about half since the beginning of the century. Ponds disappear when people use them to deposit rubbish or where they are filled in because the land is needed. Many of those remaining are damaged by land drainage, pollution, neglect and mismanagement. Woods are one of the few areas of the countryside where extensive areas of shallow water wetlands and wet ground still exist.[8] Pond conservation should therefore be a high priority for woodland owners.

The Origin of Ponds

Natural Ponds

Some woodland ponds are of natural origin. Indeed, the presence of a pond may have been that very obstacle to cultivation that caused the area to be retained as woodland in the first place. Ponds without any sign of banks are likely to be of natural origin.

Man-Made Ponds

These were dug for watering places for cattle, for fish-farming and for duck decoys. Some originate from holes dug to extract minerals or peat, and charcoal was first made in pits. Man-made ponds can sometimes be recognized by low mounds of excavated material lying nearby. Even where ponds have resulted from the removal of material, there is usually a bank of overburden or spoil.

Moats

In the 12th and 13th centuries moats were sometimes constructed around houses – probably as status symbols. These moated houses were often located in woodlands, but in other cases woodlands may have grown up around them. Moats are easily

ENGLISH NATURE

The value of small woods for nature conservation

Dr Keith Kirby, English Nature

Small woods are an important part of the fabric of the countryside: on many farms they are the richest areas for wildlife and are essential if populations of common (and sometimes not so common) species are not to be confined in future to a few large nature reserves. English Nature therefore supports their conservation and management as part of our commitment to maintaining England's biodiversity.

Small woods include many nineteenth century game coverts and spinneys and corners of fields that scrubbed up in the early part of this century, but they also include many fragments of ancient woodland. These may not be listed as such on the Ancient Woodland Inventories prepared by the Nature Conservancy Council and its successors, which do not include woods below 2 ha, but that does not make them any the less ancient. They may contain species such as bluebells and anemones, old coppice stools, woodbanks and the other features found in larger ancient woods. They are also more likely to be made up of native trees and shrubs because there has been less economic incentive to replant small woods with conifers.

A group of small woods may be richer in species than a single wood of the same total area because the individual small woods may be spread across a wider range of soil types and site conditions than the single wood. The small woods are less likely to contain species that require dark 'forest interior' conditions, because no part of a small wood is far from the edge. However many birds, butterflies and plants use woodland edges for at least part of the time and these species will benefit from maintaining small woods, improving their management or creating new ones.

There is good evidence that a varied landscape with hedges, rough grassland patches and small woods is less of a barrier to movement of plants and animals through the countryside than an open uniform one, such as intensively farmed landscapes. In addition, if new woodland is added to the fragments that exist already, there is an immediate source of woodland species on hand to move into the newly created habitat.

Small woods are therefore valuable as a habitat in their own right, for the contribution they make to the surrounding countryside and to the movement of species through the landscape.

recognized. They are commonly four-sided, usually square or rectangular but sometimes circular or irregular. The houses they enclosed were generally made of wood and have left few traces.

Bomb Craters

At least a few woodland ponds must result from bomb craters. These are distinguished by their neatly circular shape with the displaced soil forming a mound all around.

Earth Dams

These are not uncommon in woodlands. They were often built for fire-fighting purposes, particularly where conifers were grown.

Whether a pond is natural or man-made will have little effect on the type of plant and animal communities it supports.

Pond Succession

While some natural ponds can be very stable, changing little over thousands of years, others can be relatively short lived – created and then progressively filled in with sediment over hundreds rather than thousands of years, and in some cases in just a few decades. For most types of pond a gradual succession from open water to dry land is a perfectly natural process and all stages of the succession, from open water, to silted pond, muddy hollow, waterlogged soil, and finally dry land, will be exploited by some wildlife. Newly created ponds support aquatic invertebrates and plants not found at later stages of succession. Surprisingly, perhaps, some may be rare. As ponds gradually change, so the communities they support will alter. Many species are able to move between ponds as conditions become unsuitable.

For many people the natural processes that lead to a gradual loss of open water will be seen as something undesirable. Some wet areas may seem nothing more than dark, soggy areas of rotting leaves and mud that are totally uninteresting. There may be a desire to dredge and restore such ponds in an attempt to 'improve' their conservation value, or to fill them in as worthless. Either decision can result in a serious loss of habitat.[9]

Pond Diversity

Ponds of all shapes and sizes, and depths and degrees of permanence have the potential to provide valuable habitats. The less intensively the land around a pond is used, the more likely the pond is to have a diverse plant and invertebrate community and to support uncommon species.

Stagnant Ponds

These will be colonized by a distinct range of plants and animals that are well adapted to the oxygen stresses found in smaller bodies of water.

Seasonal Ponds

Seasonal ponds are those that dry out, usually in the summer, but are nevertheless permanent features of the landscape. The animals of seasonal ponds have mechanisms to survive summer drought. Although the plants and animals that can tolerate these extreme conditions are limited, they are specialized and may include uncommon or rare species.

Shaded Ponds

The muddy edges of shaded ponds can be an important habitat for the larvae of some insects.

Shallow Ponds

Shallow ponds in long-established woodlands can support many of our rarest wetland flora and fauna. Tree trunks and branches that fall in the water become substrates for species of algae and fungi, and as the wood rots it provides egg-laying sites for dragon-flies and food for aquatic fly and beetle larvae. These sites can be of very high conservation value. Such ponds are a feature of woodlands where human disturbance is infrequent or absent. Conversely, in areas subject to intensive management (such as commercial forestry plantations), heavily mechanized operations can churn up soil surfaces leaving deep ruts and damaged drainage systems. Surface runoff can pollute ponds, allowing only the most common and resilient species to thrive. Such ponds will generally be of low conservation value.

Pond Restoration

Before making any decision on the restoration of a pond it is essential to assess its value for wildlife. Find out as much as possible about the pond, including the following.

- How, when and why was it made?
- Has it been managed in the past?

- Is it fed from a spring, or from surface or groundwater?
- What is the quality of the water?
- How much water is held?
- What are the depths of its various levels?
- How much sediment exists?
- How fast is sediment building up?

A survey of the flora and fauna will be needed to assess the status of the wetland area. This is easier said than done. The complexity of pond habitat makes it very difficult, though not impossible, to survey and assess all the groups of plants and animals present – and they may vary from year to year. It will require the services of a specialist. Where a survey is not feasible, it will probably be better to do nothing other than to create a new pond if a suitable site is available nearby.

Once the area is better understood, proposals for management and protection can be considered. However, there is still insufficient information about the effects of management on ponds to make anything other than very general recommendations. One of the most significant myths of pond management is that drying out is disastrous for wildlife.[10] Temporary ponds require just as much attention, and protection, as more permanent sites.[11] Recent opinion has brought into question whether some traditional ideas of pond management are actually beneficial to their conservation. Many guides advise that ponds must not become choked with vegetation. However, it is probably better to have too many plants in a pond than too few. Open water is a dangerous place for most pond animals and the majority live in the more protected habitats associated with plants or sediments. For many species, plants are vital at one or more stages of their life-cycle. A range of plants is important to encourage a variety of animals, and different densities of plants are known to be important for encouraging distinct animal communities.

Pondside Trees

There is also uncertainty about how best to manage trees around ponds. Much advice has concentrated on the supposedly adverse effects of trees, particularly the dark and dank conditions that they can create. But shaded ponds are just as likely to support rare and uncommon species as are more open sites. It is thus advisable to retain shaded ponds wherever possible, though where they are common, some can be carefully opened up by the removal of a few edge trees to give a good balance between sunny and shaded areas. It is probably better to thin out marginal trees rather than to remove them wholesale. Trees should be felled away from the pond to avoid damage to marginal vegetation. Where dredging is being considered:

- Avoid drastic change as far as possible.[12]
- Only dredge as a last resort, and then only in ponds that have been periodically cleared out.
- If dredging is considered essential, restrict it to part of the pond only.
- Reduce but do not remove the sediment completely.
- Save clumps of vegetation to be replanted around the margins to reestablish important habitat.
- Dump sediment only on land of low conservation value.
- Check whether the pond is clay-lined – or the water may be lost and repair work could prove costly.

DESIGNING AND SITING A NEW POND

Ponds are created for a variety of purposes and all will support wildlife. However, this section deals primarily with the creation of ponds in order to specifically encourage wildlife. In deciding on the site for the new pond, first assess the vegetation to ensure that the existing wildlife value is not greater than that which will be generated by the pond. There are two types of ponds:

- below ground – where the water is retained in an excavation entirely below the natural ground level;
- above ground – where the water is retained by dams and embankments; above ground constructions should not be attempted without the advice of a chartered civil engineer.

There are three possible sources of water:

- surface water, which relies on land drainage (check the size of the catchment area);
- groundwater, which depends on the soil type and geology (dig a few test pits and monitor for at least a year);
- springs, which are usually a reliable source of water for ponds.

Of the three constituents of soil (sand, silt and clay), clay is the most important ingredient for impermeability. Investigate the soil by referring to soil maps or by using a soil auger over the whole site. In planning the pond:

- Maximize the margin by designing a convoluted edge.
- Aim for a progression of depths from very shallow to a couple of metres.
- Vary the gradient of the edge zone, ensuring some very gentle slopes so that animals have easy access to the water.
- Aim to provide a range of pondside light conditions from open areas with grasses and herbs, through patches of scrub, to mature trees.
- Keep the southern and eastern sides free of tall trees to increase the amount of sunlight reaching the pond.
- Retain as many native British tree and shrub species as possible to provide the greatest number of insects, birds and animals.
- If trees need to be felled to make way for the pond, keep the tree line at least 20 metres away from the proposed margin.
- Unless they are relatively small, do not leave isolated trees standing as they will be prone to windthrow – it is better to plant marginal trees where needed.
- Remember that trees take up large volumes of water during the growing season and this may lower the watertable locally.
- If feasible, link the pond with the existing woodland ride system.
- Consider constructing a pondside hide from which wildlife may be studied.

Islands

Islands provide havens for wildfowl but foxes, stoats and weasels can cross small areas of water, so islands are of only limited value where small ponds are concerned. Islands should be near water level. They can be created naturally by leaving an area untouched during excavation. Floating islands can also be made. They require anchoring and should incorporate a gently sloping ramp to give animals easy access to and from water.

As an alternative to digging an isolated pond, if space is available consider the construction of a wetland complex incorporating temporary pools, muddy shallows and marshy areas. Such a scheme would lend itself to the construction of a number of hides, screens and seats from which to watch wildlife.

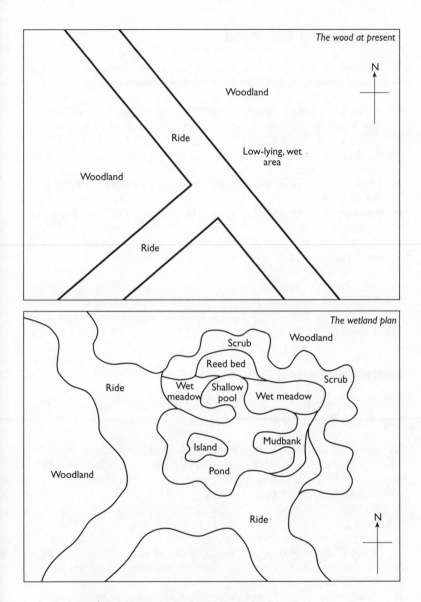

Figure 5.1 *Planning a Wetland Complex*

Constructing the Pond

The information provided in this section is for guidance only and is not intended to substitute for professional advice.

- Draw up a simple plan showing the outline of the pond and the locations of islands, spits, bays, shallows and deep water.
- Once a suitable location has been chosen, peg out the position of the pond and calculate the approximate volume of earth to be removed.
- Identify an area of low conservation value where the spoil will be spread.
- Contact a plant operator skilled in the construction of ponds valuable for wildlife.
- Ask to see examples of similar work that the contractor has undertaken.

The best time to construct a pond from a conservation standpoint is winter. If the water in the trial pits has remained high, there will be no need to line the pond because it will fill naturally with groundwater.

Lining Materials

If the watertable is too low, artificial liners or imported clay puddling will be necessary. With regard to linings:

- Butyl rubber heavy-duty sheeting is expensive but is generally guaranteed for 50 years and is suitable for small ponds.
- Heavy-duty polythene is cheaper but is more difficult to handle because it is more rigid.
- Linings require careful handling and laying on a stone-free base.
- They should be buried beneath a protective layer of soil to prevent damage through livestock trampling, vandalism, etc.
- In stony or flinty soils, some linings will require a layer of sand beneath them
- Penetration of linings by weeds can also be a problem and weedy soils should be treated first.

Some materials are weakened by ultra-violet light and they require a covering of soil to protect them, in which case slopes of a ratio of not more than 4:1 are advised to prevent the soil from sliding to the base of the pond. The soil is also needed as a rooting medium for aquatic or marginal vegetation. Large plant species whose roots may penetrate the lining must be avoided.

A new pond will very quickly acquire an interesting natural community of plants and animals, but species can be brought in to help colonize the pond. Some water and sediment from an established pond will introduce small species such as snails and insect larvae. It should not be necessary to introduce amphibian species to the pond; if the habitat is suitable they will find their own way.

Algal blooms are a common phenomenon in newly created ponds. The problem usually takes care of itself in a short time once plant and animal communities become balanced systems. If the problem persists, barley straw placed in net bags and allowed to float seems to inhibit the growth of algae. Use about ten grammes of barley straw per cubic metre of water and apply twice a year in spring and autumn. Chemical preparations should not be necessary. It is useful to keep a diary and photographic records of the work's progress.

Other Considerations

- Avoid the temptation to link land drains to ponds where there is any risk of enrichment due to fertilizer application.
- In certain circumstances a licence may be necessary for impounding water, abstracting water, making a reservoir, draining land or stocking fish. Check with the Environment Agency before starting any pond work.
- Prior notification of any plans to build a pond is required by all local authorities. Planning permission will be necessary in certain circumstances.
- Under the Wildlife and Countryside Act it is illegal to move certain protected species of animals from their native habitat.
- It is illegal to release non-native waterfowl into the wild unless their wings are pinioned or clipped so that they cannot fly. English Nature should be consulted for further information.
- Whatever the size of the pond, owners should always be aware of the dangers associated with water. Warning notices or fences may be needed near ponds where visitors are anticipated.
- Grants may be available to help with the creation and management of ponds. For details contact your local authority.

6

Recreation

Demand for countryside recreation has increased substantially over the last couple of decades and is likely to go on increasing. Woodland visits give great pleasure – people love the outdoor life, the trees, the wildflowers and the wildlife – and woodlands are ideally placed to offer a fascinating range of outdoor interests.

PRIVATE RECREATION

While many owners of small woods are disinclined to grant access to the public, it is an option worth considering, depending on circumstances and personal preferences.

As with any other objective, before planning for recreational improvements the first essential is to identify the most important features in the wood. It may then be possible to link most, or all, of the features by laying out a meandering path. Depending on circumstances, the trail should follow a circuitous route, weaving this way and that, making the most of whatever space is available. The following features are just some that could add interest to a walk:

■ glades;
■ wet areas;
■ ponds;
■ streams;
■ old wells;
■ hollows;

- rocky outcrops;
- veteran, hollow or dead trees;
- old pollards;
- large tree stumps;
- boundary banks;
- old walls;
- derelict buildings;
- old charcoal burning sites;
- as many different tree and shrub species as possible.

Badger setts also add interest. They are protected by law but a path passing within sighting distance should not disturb them.

Consider felling occasional wedge-shaped sections of trees where an attractive vista over surrounding countryside would add diversity. Ensure paths are wide enough for two people to walk abreast and move brash and other debris well back from the edge where it can look unsightly. Where interesting features are limited:

- Create new glades.
- Use logs as wayside seats.
- Construct rustic bridges over drains and streams.
- Place wooden walkways across wet areas.
- Make a picnic site.
- Build a barbecue from local stone.
- Instal a wildlife observation hide.
- Put up a 'high seat' for wildlife observation.
- Construct a new pond with marshy surrounds.
- Widen rides.
- Create some bare areas under large trees where children can run about.
- Plant willow cuttings in a semicircle at some point on the route to grow into a living bower with a seat.
- Plant a dense clump of holly, yew or box to create a spooky area of dark tunnels where children can play games and make dens.
- Thin trees to provide a variety of light and shade conditions.
- Put up a tree house and a 'tarzan' swing for children.
- For the truly adventurous (and the well insured), consider constructing a wooden walkway, way above ground level, through the branches of the trees.

To help keep the woodland screened from the outside, evergreen trees and shrubs can be planted around the boundary. Where public rights of way exist, unintended trespass can be reduced by adequate waymarking and the provision of gates and stiles. Public paths should always be kept clear of undergrowth and low branches cut well back (see Chapter 1).

To help keep children amused during school holidays, organize some woodland projects. Let them:

- Learn to identify trees and shrubs. NAME TAGS
- Collect and sow some tree seeds.
- Plant a mini arboretum.
- Measure and keep a record of tree heights and girths.
- Make a 'leaf library' of preserved leaves.
- Make a collage of leaves.
- Make bark rubbings and leaf prints.
- Construct and site nest boxes and record usage.
- Organize a treasure hunt.
- Map and survey the wood.
- Study and record animals, birds, butterflies, etc.
- Make plaster casts of animal tracks.
- Start a woodland museum with examples of tree bark, galls, cones, etc.

PUBLIC RECREATION

The extent to which opportunities can be provided for different forms of public recreation will depend on the location of the wood relative to centres of population and tourist routes, as well as its structure and size. Generally, the larger the woodland the more capable it will be of accommodating high numbers of visitors without significant damage or disturbance.

Small woods are more restricted in scope for recreation. They are generally best suited to informal use by their owners and by local people, though some may be suitable for a number of low-key commercial activities. A wider range of activities becomes possible where several small woods fall to the same ownership and where a route can be introduced to link them.

Most visitors do not visit woods with any particular activity in mind. Many stay close to their cars and some do not leave them. Those who walk rarely go far, though they may be persuaded to follow a waymarked trail. Assessment of the site should aim to establish if anything vulnerable is present, so that informed decisions can be made to minimize damage. The likely impact of different options should be carefully thought out to ensure that access does not conflict with operational work, or lower environmental and conservation values. Where circumstances permit, specially designed access facilities for disabled visitors are well worth considering, while good public relations can be fostered by involving local people in any decision making process. Such measures are likely to attract a wider range of management grants.

The Waymarked Trail

The most basic provision for public access is the footpath. This can mean creating paths where they do not already exist or possibly upgrading those already in place. A wood of ten hectares can accommodate a path about three-kilometres long; a wood half that size could still absorb a one-kilometre trail.

A leaflet describing points of interest along the trail will greatly add to enjoyment and can help the public to understand the work of the private woodland owner. At the starting point, an information board will be needed. This should briefly describe the length of the walk and the conditions to be expected. The layout of the walk in relation to the wood can be shown as a simple plan. The board should be positioned so that it depicts the lie of the land directly in front of the observer. It needs to be low enough for children to read it. Boards erected at an angle, lectern style, allowing the observer a view of the land beyond can be very effective.

Effective waymarking is a skilled business. A circuitous route is best, but a decision must be taken at the outset as to whether it is to be a oneway (directional) or an either-way (go as you please) system. Many more markers will be needed for signing an either-way trail. One-way systems are useful where:

- arduous sections (for example, uneven ground) are best negotiated early in the walk rather than when visitors are more likely to be tired;
- some particularly good view can best be appreciated looking ahead rather than over the shoulder;
- the walk has an educational theme that needs to be presented in an orderly sequence.

The first marker post on a directional walk must be very prominent; to avoid any risk of confusion, ensure the last marker on the trail is not visible from the starting point. Acute changes of direction must be avoided. Walkers engaged in conversation will often miss them – even where conspicuously sited. Where unavoidable, some rustic poles can be used to guide walkers. In woodlands large enough to accommodate more than one walk, it is preferable to lay out each as a separate route rather than having a network with multicoloured marker systems.

Ensure that there are sufficient markers. One will be needed just beyond each change in direction and, on straight sections,

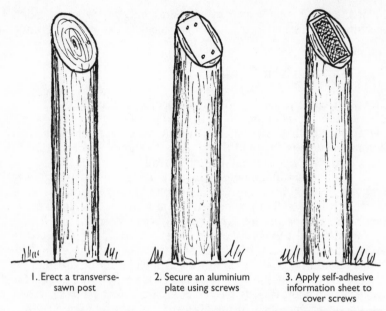

1. Erect a transverse-sawn post

2. Secure an aluminium plate using screws

3. Apply self-adhesive information sheet to cover screws

Figure 6.1 *Design for a Waymarker Post*

confirmatory markers are required at 400- to 500-metre intervals, even where the route is obvious. If there are too few, less intrepid visitors will be inclined to turn back. Frequent inspection and maintenance of all woodland walk equipment is vital.

Anyone with an approved woodland grant scheme and who provides public access can obtain from the Forestry Authority a free 'Walkers Welcome' package consisting of a variety of signs for use at woodland entrances and discs for display at appropriate places along routes through the woodland. They will also provide, free of charge, a range of signs to explain the reasons for various woodland operations such as ride widening, coppicing, pollarding, felling and thinning. They also hold periodic Woodland Access and Recreational Planning courses. For details contact the nearest Forestry Authority office.

In the absence of ready-made signs, homemade notices will serve almost as well. Run off on a word processor and encapsulated in heavy-duty plastic, they can be affixed to convenient gates, fences or posts.

COMMERCIAL RECREATION

The public is willing to pay for leisure provided the quality of the experience and the enjoyment are rated as good value. In some circumstances, even small woodlands can lend themselves to management for commercial recreational use, bringing potential economic benefits for the owner and employment opportunities for local people. By linking together separate woods using cross-country paths, the value of the enterprise can be greatly enhanced.

Owners can charge for a range of facilities, goods and services and may be eligible for certain grants. The most basic option lies in levying a charge for entry. Subscribers are provided with a gate key, a set of conditions and perhaps entry to a car park. Other options may include:

- charging for car parking;
- selling refreshments;
- selling books;
- selling trail guides;
- selling educational packs;
- selling souvenirs;
- selling charcoal and firewood;
- selling other woodland produce;
- selling trees and shrubs;
- selling Christmas trees;
- ice cream concessions;
- charging for school visits.

Specialist activities worthy of consideration include:

- interpretive nature trail;
- guided walks;
- adventure trail;
- adventure playground;
- fitness trail;
- activity trail;
- orienteering;
- military training;
- horse riding;
- sculpture trail;
- bird watching events (dawn chorus);
- educational facilities;
- tree work demonstrations;
- endurance course;
- woodland museum;

- arboretum;
- planting of 'in memoriam' trees;
- war games;
- paint ball games;
- organized camps;
- woodland craft workshops;
- clay pigeon shooting;
- pheasant shooting;
- deer stalking;
- fishing;
- pony trekking;
- archery;
- mountain bike hire;
- camping;
- barbecue hire;
- log cabin hire ;
- children's playground;
- caravanning;
- pet's cemetery.

Rarely will a small woodland be able to accommodate more than one or two activities at any one time. The need for planning permission will depend on the chosen development – check with the local planning authority if in doubt. Liaison and consultation with others likely to be affected by the development may also be appropriate. Local people who may have used the wood for many years could be offered free passes or season tickets.

In almost all cases when planning for recreation, an annotated map is indispensable. Provision must be made for any constraints such as sensitive wildlife habitats or archaeological sites. The possibility of providing access for less-able visitors needs careful thought. The provision of toilet and washing facilities may need to be evaluated. Owners should be mindful that they have a duty of care for visitors to their areas so aspects of safety, liability and insurance must be considered.

Some market research will almost certainly be needed. An estimate of the number of visitors likely to be attracted to the site should be compared with an estimate of the number the woodland can accommodate without compromising other objectives. Thought must be given to whether, and how, the woodland is to be publicized and how visitor numbers are to be limited should a site become so popular as to lead to overuse.

There is much else to think about:

IMPORTANT! (handwritten)

- Will unwanted tracks be created in the wood?
- Will wildlife be disturbed or flora trampled?
- Will the intrinsic character of the area change?
- Will the commercial nature of the activity attract other unwanted enterprises (for example, ice cream vans)?
- How will litter be disposed of?
- Is the site liable to vandalism or theft?
- Are dogs to be allowed in and if so will the mess they make inconvenience walkers? *- compost poo bins! + show on map!* (handwritten)

The assessment should conclude with a cost-benefit analysis in which the capital cost, the running costs, the potential for grants and the predicted returns are all considered. With the enterprise up and running, the owner will want to know if the development is succeeding. Progress can be monitored by keeping records of visitor numbers and the length of time they stay. Questionnaires can provide useful information at relatively low cost – they can be used to judge the level of satisfaction and to plan for future changes.

Sporting

While the issue of country sports is a matter for individual conscience, there is little doubt that they have brought considerable environmental and economic benefits to the countryside.

Game shooting is in great demand and an increasing number of woodland owners and managers recognize the potential of managing their woodlands for sporting use. Where large woodlands are concerned, sporting rents can play an important part in woodland economics, adding considerably to the capital value of the property. Small woods, on the other hand, are more likely to provide informal sport for the owner, though rights can still be let to individuals or to a shooting syndicate for some modest sum provided the wood offers suitable habitat. Some owners will allow the shooting tenant to manage the wood on his or her behalf, thus securing a regular income with no expense. An increasing number of owners let rough shooting by the day, with charges varying according to the quality of the sport. The secret is to create the right conditions to allow game species to increase. Such conditions will also improve prospects for many other species.

Those who would like to shoot but do not themselves own land should make enquiries of land agents, sporting agents, the local press, sporting magazines and forestry companies. It is necessary to obtain permission (preferably in writing) from the landowner, the holder of the shooting rights or some other authorized person before one may legally take a gun on to private property. It is also essential to ascertain what species are to be included and excluded in the terms of the shooting agreement.

The law is quite specific as to what may be shot and when. Under the Wildlife and Countryside Act, all birds are protected except those that are specifically excluded from protection in the legislation. Amongst the birds that can be shot at any time are jackdaws, jays, magpies, wood pigeons, rooks and carrion crows. Others, including pheasant, partridge, woodcock and snipe, may be shot only in the specified season.

Before holding a gun, a shotgun licence must be obtained from the police, and if the plan is to shoot pheasant, partridge, grouse, blackgame, ptarmigan, capercaillie, woodcock, snipe, or hares then a game licence must be obtained from the local post office. There are strict laws covering deer shooting and the use of firearms; these differ between Scotland, England and Wales.

PHEASANT SHOOTING

For pheasants to thrive, the following conditions are needed:

- plenty of woodland edges;
- plenty of cover and shelter;
- ample food;
- protection from predators.

Blocks of small woodland between one and three hectares in area are ideal because pheasants spend most of their time within a few metres of open ground. Over-large sections make it difficult to get birds out and when they do emerge they are often exhausted. Grain is usually provided but winter pheasant density can be influenced by natural food availability.

The presence of ground cover and a perimeter hedge improves shelter and warmth. The canopy should not be so dense as to suppress all ground vegetation because this can create draughty, cold conditions. Routine thinnings need to be on the heavy side to maintain the right intensity of light. Drainage should never be neglected; a damp wood can prove fatal to young birds.

Many owners are discouraged from clearfelling and replanting because of the disturbance caused, though in some cases (for example, where a conifer crop is approaching maturity) few alternatives exist. The Game Conservancy recommends a less drastic approach that can significantly improve sporting potential at a reasonable cost. It involves clearing small areas of trees to form 'skylights' in the woodland canopy and, where light conditions are suitable, replanting with a scatter of trees and shrubs, protected with treeshelters. A good distribution of skylights can have a dramatic effect on the ground vegetation in a couple of years. The size of the skylights is determined by the height of the trees. About 15 metres in diameter is usually the right size to start in older broadleaved crops. If the skylights are too large initially, this can allow an excessively dense undergrowth to develop. If growth of the newly planted trees is being unduly affected by shade, the skylights can be enlarged.

When choosing species for replanting it is wise to avoid sycamore as it casts a heavy shade and renders the surrounding woodland floor bare. Beech provides mast, a useful pheasant food, but it too casts a dense shade so should be used sparingly. Oak, ash, wild cherry, rowan and birch are more likely to permit the regrowth of bramble, hawthorn, holly, blackthorn, dogwood, elder, spindle and privet, and the inclusion of a few conifers (such as larch and Scots pine) is useful for roosting. Managed coppice can provide the most attractive breeding and holding ground for pheasants. Ideally, it should be grown in small blocks of different ages to provide a sequence of open, sunny patches, low cover and more dense areas for warmth and shelter. Large areas of equal age can present problems for beaters.

The management of rides is important for access and feeding. Left unmanaged, narrow rides become nothing more than gloomy tunnels beneath the trees and wide ones become overgrown with a tangle of vegetation. Maintenance is necessary to keep rides open, warm and sunny. Regular mowing, once or twice a year, is advisable and an uncut strip of about three metres can be left either side to provide cover and nesting sites. Ideally, a wood should have a main central ride at least 20-metres wide and several narrower ones. Straight rides that open directly at the edge of a wood let in the wind. They can be improved by blocking them off with some appropriate tree and shrub planting. A new access can then be cut through at an angle. Even with the ends blocked, long, straight rides can be cold and draughty. A few groups of trees planted on the sides of the ride can serve well as baffles and will provide additional cover.

The ideal pheasant wood should have a regularly trimmed outer hedge. Many have a row of dense evergreen trees (such as western red cedar) to keep the wood warm. Patches of flushing cover of low shrubs and young trees will encourage the birds to rise when the wood is beaten. Pheasants rise at an angle of about 25 to 30 degrees. If they are forced to rise at any steeper angle they soon lose power. Flushing points are normally spaced at intervals through the wood, taking advantage of any rising ground. Beyond the flushing points a few rows of tall trees should be retained to force the birds to keep rising to a good height above the guns. A final flushing point of densely maintained coppice should be positioned at the end of the wood to pick up any remaining birds.

In conclusion, gamekeepers: please take note. Correctly planned and promptly tackled, woodland operations should not cause great disruption to the sporting calendar.

ROUGH SHOOTING

Whereas the pheasant shooter waits at a prearranged point for the quarry to fly past the gun, the rough shooter is primarily a hunter of game. Though there are no hard and fast rules to distinguish one form of shooting from the other, there is a degree of flexibility and informality associated with the rough shoot that is absent in a driven shoot. Rough shooting involves much tactical manoeuvring and an ability to adapt to changing circumstances.

It is difficult to say what might be regarded as the minimum area that could be described as a useful rough shoot. Much will depend on the type of ground and what game is to be found. Even the smallest wood can form the basis of a shoot provided the potential is there. Broadleaved woods and mixtures of broadleaves and conifers generally offer a richer environment for game than pure conifer blocks. They provide better availability of food, shelter and nesting cover.

Blocks of conifers, while young, can provide moderately productive sport but they become progressively of less value as the canopy closes and the light is excluded. Few game birds or animals will inhabit them at that stage in their development. Conditions can be expected to improve towards the end of a conifer rotation, provided the crop has been regularly thinned. As the light intensity strengthens, so the ground vegetation will provide better habitat for game.

When replanting felled areas, a varied species composition with a good intermix of broadleaves will provide the right conditions to attract game and will help prevent resident birds from straying. Rough patches of land should be retained in the wood itself or in adjacent field corners. Other changes that may bring significant benefits include:

- creating new rides;
- widening and scalloping existing rides;
- redesigning boundaries to produce sinuous outlines;
- opening up glades;
- planting shrubs for cover;
- planting strips of cover crops in adjacent fields.

DEER STALKING

Deer stalking can provide a useful source of income but it is best suited to estates with large areas of woodland. In smaller woods, deer management is usually not so much a sport as a necessity.

Where deer are concerned, there is no requirement for improving habitat to increase numbers. Deer populations, particularly in southern Britain, have been increasing relentlessly for some years. There is more likely a need to modify woodland design to permit effective control – while many people will welcome this expansion to the woodland fauna, damage by deer (which is principally to young trees) has become a serious problem for woodland owners (see Chapter 3).

The preferred habitat of breeding roe and muntjac adults is young woodland with dense cover. Red, fallow and sika prefer older or mixed-age woodlands. To monitor and control deer, they need to be seen. Grassy rides and sheltered glades are a major feature of control. Glades, a quarter to half a hectare in area, should be developed in areas used by deer. Their effectiveness can be enhanced by supplementary feeding. Rides should be mown at least every other year to maintain good browse. 'High seats' can be constructed overlooking ground where thicket-stage crops meet open grassy areas. They can be a most productive method of deer stalking and they are certainly the safest.[1] Also useful are portable lightweight tubular metal seats.[2] These can be moved and positioned exactly where they will be most effective.

To prevent deer populations from increasing, around 20 to 25 per cent of the adults may need to be shot each year. Culling should concentrate on mature females. Warning signs should be erected at any points of public entry into woods where culling takes place and, to avoid concern, these could briefly explain the need for culling.

Owners cannot always justify the expense of employing a professional stalker to maintain numbers at an acceptable level and so they may choose to carry out culling themselves. Shooting is the only permissible method of culling deer and this must be done humanely and within the relevant Acts of Parliament. Landowners normally have the right to shoot wild deer on their land. In Scotland tenants have the right to shoot deer that are causing damage to enclosed woodlands, crops or grassland. In England and Wales tenants' rights are more variable: tenants should check their leases and seek advice. Only rifles of specified calibre and muzzle energy can be used, although under certain circumstances a shotgun may be permitted. Night shooting is illegal in England and Wales.

Legal close seasons when deer may not normally be shot vary between species, sexes and countries. The work calls for training to a high standard if deer populations are to be managed humanely and effectively. Haphazard amateur efforts can compound the problem (for example, where insufficient females are culled). Owners and managers with little experience of deer should seek expert advice and help with culling. Several bodies give training and award certificates to stalkers. Local offices of the Forestry Authority, Agricultural Development and Advisory Service (in England and Wales) and the British Deer Society may be able to advise on qualified stalkers in particular areas.

WILDFOWL SHOOTING

Ducks and geese will often use woodland ponds and streams, thus increasing sporting potential. Conditions that will favour wildfowl include:

- water that is clearly visible from overhead;
- clear flight paths that extend well back into the wood;
- a sinuous shoreline with gently sloping margins (to encourage the growth of aquatic vegetation and to allow young wildfowl to walk ashore without difficulty);
- some stretches of deeper water for bottom-feeding species;
- sheltered banks where wildfowl can rest and preen.

Pools should not be shot too frequently – about once a fortnight is considered ideal – and there needs to be a sensible bag-limit at any one shoot. Wildfowl shooting should never take place without a trained retrieving dog on hand.

FISHING

Woodland streams and ponds are attractive to anglers only where they can cast without obstruction from trees and shrubs. Management to facilitate stream fishing includes:

- maintaining intermittent groups of trees such as birch, aspen, alder and willow (to create some lengths of lightly shaded water between more open stretches);
- the provision of natural vegetation on the stream side;
- the creation of pools (to check the flow and to provide holding water);

- the avoidance of any draining work from mid-October to mid-May (when eggs and young may be affected by silt).

Management to facilitate pond fishing includes:

- maintaining fish stocks at an adequate level;
- the provision of deep water for overwintering fish;
- ensuring a sensible balance between weed removal (to provide good angling conditions) and the provision of adequate plants (to oxygenate the water and to provide cover and spawning habitat);
- the maintenance of occasional groups of waterside trees (as a source of insect food dropping onto the water);
- sufficient natural vegetation to screen anglers.

FOX HUNTING

Foxes depend on areas of woodland for cover. Management to facilitate fox hunting includes promoting movement of the hunt through the wood. A dense wood is unsatisfactory for hunting. The provision of artificial earths to encourage foxes to breed in woodlands is to be deplored – but is not unknown.

SPORTING ORGANIZATIONS

The Game Conservancy

The Game Conservancy operates an advisory service that gives general advice in the form of books, pamphlets and research results, as well as advice specific to an individual shoot by arranging a visit from one of their field officers. It also organizes courses for shoot managers and keepers and produces a code of good shooting practice.

The British Association of Shooting and Conservation

This association has produced a series of codes of practice to guide newcomers to the sport, and has introduced a proficiency award scheme that covers both the theory and practice of good, safe shooting.

The British Field Sports Society

This society represents the interests of all sportsmen and women who follow or participate in field sports, be it shooting, hunting, coursing, fishing, falconry or stalking.

The British Deer Society

The British Deer Society provides helpful facilities and an advisory service.

Shelter

Exposure to wind can be a problem in almost any part of the country. Where they are correctly oriented, well-structured and adequately maintained, woodland windbreaks can provide:

- effective shelter for crops, livestock and buildings;
- help in preventing soil erosion;
- traps to catch drifting snow;
- sporting cover;
- timber and wood products;
- improved landscape (provided they are related to topography);
- wildlife habitats;
- a more pleasant environment in which to live and work.

SHELTERBELT STRUCTURE

Very dense woodlands create a barrier that deflects wind over the tops of trees, resulting in a calm area in the immediate lee of the wood but considerable turbulence further out where the wind meets the ground once more. This type of wood can be suitable for sheltering livestock – though owners must understand that there are drawbacks in allowing livestock free access within woods (see Chapter 3).

More opengrown woods reduce wind velocity by acting as a buffer to the main airstream, creating a longer zone of shelter and less turbulence. The most effective windbreaks are evenly permeable from top to bottom, with a windward edge that presents a

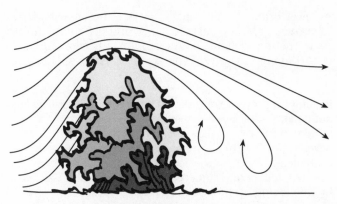

1. Dense barrier gives short zone of shelter

2. Permeable barrier gives more effective shelter

Figure 8.1 *Shelterbelt Structure*

vertical face to the wind. Long, narrow woods are best – a width of between 15 metres and 30 metres, and a length at least 12 times the height will give good shelter for a distance of up to 25 times the height of the trees. A graded, sloping edge, as recommended for good conservation management, will allow the wind to flow smoothly over the canopy with little reduction in speed.

RESTORING NEGLECTED SHELTERBELTS

Left unmanaged, shelterbelts become ineffective. Lack of thinning allows lower branches and smaller trees to become shaded and to die. The trees grow too close together and are unable to develop adequate rooting systems, so the crop may become unstable and liable to damage by wind. The restoration of shelterbelts can involve:

■ fencing (to prevent trespass by livestock and other browsing animals);
■ regular thinning (to allow trees more room in which to develop firm rooting systems and well-furnished side branches);
■ selective felling and underplanting (to help maintain structure by thickening up the lower storey);
■ phased clearfelling and replanting (to provide continuity).

In shelterbelts where timber production is also an important objective, selected trees can still be brashed or high pruned to improve timber quality. These operations can be used to regulate permeability. Planting a new shelterbelt alongside the existing one is a way of maintaining continuity. Once the new belt has begun to provide shelter, the old one can be cut down and replanted. The two blocks can then be alternately felled and replanted at appropriate intervals.

Shelterbelt replanting calls for deep-rooted, wind-firm trees capable of growing quickly on the site. Correct choice of species is crucial (see Chapter 3). Much will depend on soil type, degree of exposure and the climatic conditions prevailing at the site. A useful mixture might include two or three deciduous species such as:

■ oak;
■ beech;
■ hornbeam;
■ ash;
■ lime;
■ wild cherry;
■ grey alder.

To this could be added two or three species of smaller stature, such as:

■ field maple;
■ crab apple;
■ birch;
■ rowan;

■   whitebeam.

It is worth including one or two evergreen species, such as:

■   Corsican pine;
■   Scots pine;
■   yew.

Finally, to provide an understorey, plant a few shrubs such as:

■   hazel;
■   blackthorn;
■   hawthorn;
■   holly;
■   privet.

Such a mixture should create a well-structured shelterbelt capable of providing good protection throughout the year. Planting can be at conventional spacing. There is nothing to be gained by a closer than normal spacing. This will only precipitate the need for earlier and heavier thinnings.

Appendix 1

Recognizing Ancient Woodlands

The term ancient woodland is used to describe a site that has been continuously wooded for at least the last 400 years. The year AD1600 is used as a convenient starting point; until then, tree planting was relatively uncommon, while afterwards it became increasingly popular. Reliable maps, too, begin from about that time.

Most of the woods that existed in 1600 were already old by that date and some may be the direct descendants of the original prehistoric woodland that covered large parts of Britain 7000 years ago. None of them are truly natural woodlands, however, since they will all have been influenced by human activity to some extent. Nevertheless, they are of inestimable value. They have provided a refuge and a wide range of habitats for a great variety of plants and animals over the centuries, while all around them the landscape has been subject to change. Many woodland species depend entirely for their survival on the continued existence of these woodlands.

A provisional list of ancient woodlands for each county was introduced in 1981 by the former Nature Conservancy Council (NCC) from maps and historical records and some surveys. The NCC's successor bodies (English Nature, Scottish Natural Heritage and the Countryside Council for Wales) have continued the work. An owner may consult the appropriate body to check whether a particular woodland is on the inventory. Unfortunately, the inventory identifies only ancient woodland that is over two hectares in size. There are, of course, many smaller woodlands that are of ancient origin. At present, checking the significance of these must rely on local enterprise. Sometimes a local conservation group can assist but even when assistance is not available, identifying ancient woodland should not be too difficult, even for the layman, and there is considerable satisfaction to be gained from the exercise: to discover a remnant of ancient woodland is akin to discovering an ancient monument.

Even where ancient woodland sites have been replanted they may still retain ground flora and other species of particular conservation value. The aim should be to identify those species in order to preserve and enhance them.

WOODLAND NAMES

A good starting point is to consider the wood's name (if it has one). Typically, ancient woodlands have names such as:

- wood;
- grove;
- hanger;
- coppice;
- copse;
- lea.

Or they may incorporate tree names, such as:

- oak;
- ash;
- nut;
- hazel;
- beech;
- withy (willow);
- aller (alder).

Alternatively, they may have names suggesting old industry:

- tannery;
- lime kiln;
- brick kiln.

A wood may not be ancient if it has no name or has one of the following names:

- covert;

- spinney;
- plantation;
- brake;
- common;
- furze;
- gorse;
- withy;
- piece;
- field;
- lot;
- warren.

OLD MAPS

It is worth checking to see if the wood was present within any of its recognizable boundaries on early maps. The earliest available county maps differ between counties. The first edition OS maps (1800–1818) may be seen in county archives or libraries. They are also available from the publishers David & Charles. Typically, ancient woodlands:

- have sinuous, irregular boundaries;
- do not conform with the surrounding field pattern;
- are often adjacent to commons;
- are frequently on or near parish boundaries.

They may also be found:

- on steep slopes;
- on the sides of gulleys;
- along stream valleys;
- adjacent to common land or heath.

On the other hand, where a wood has straight, angular boundaries and conforms to the surrounding field pattern, or if it is located well away from the parish boundary, it suggests more recent origins.

SURVEY WORK

The features to look for inside the wood are those which imply use as a woodland since early times:

- ditches and earth boundary banks (often absent in upland woods);
- internal banks;
- charcoal-burning sites;
- sawpits;
- pillow mounds (medieval rabbit warrens);
- old kilns;
- old coppice stools;
- old pollards;
- other old trees;
- standing and fallen dead wood;
- species that are rare or of local distribution;
- sinuous streams;
- natural-looking ponds.

Secondary woodlands, on the other hand, may contain clear indications of land previously used for farming, namely:

- internal field walls;
- evidence of old hedges;
- ridge and furrow (but where trees colonized the site before AD1600, this will be regarded as ancient).

The presence of trees that are all of one age is not a reliable indicator of woodland status – it is more likely a function of management, and the site itself may or may not be ancient.

ANCIENT WOODLAND INDICATOR SPECIES

One of the most obvious differences between ancient and more recent woodlands is the composition of the vegetation. Some plant species are strongly associated with ancient woodlands. This does not mean that the presence of a few indicator species necessarily signifies an ancient woodland. Indicator species can occur in other places, and an indicator in one area may not be found in another due to variations in soil type and climate. Only where there are significant populations of indicator species and where there is also some historical evidence can reliable inferences be drawn. Native trees and shrubs provide useful pointers. It is generally accepted that in Britain today there are 33 to 38 native tree species (depending on the definition of a tree) and this includes three conifers.[1] In addition, there are a small number of native woody shrubs. Several species are by no means native all over the country; for example, small-leaved lime and Scots pine are confined to limited areas.

Lichens

Lichens, those moss-like fragments of vegetation found growing on trees and other surfaces, can also reveal clues about a woodland's past. Certain species and communities are associated with ancient trees.[2] They may persist on individual trees for many years, even when the rest of the wood has been felled. Caution is needed in interpreting evidence, however: the natural distribution of lichens has been severely modified by atmospheric pollution. Clean air is essential to most lichen species, so they soon die out or take on a desiccated appearance near large industrial centres. Agricultural sprays and fertilizer dust are also believed to affect distribution locally.

Nowadays, only in parts of Kent, Sussex, Hants, Wilts, Dorset, Devon, Cornwall, Somerset, Hereford, Northumberland and Cumbria in England; parts of Dyfed, Gwynedd, and Powys in Wales; and much of Scotland, are atmospheric conditions sufficiently pure to permit the existence of near-normal lichen populations. Large areas around the industrialized and heavily built-up regions of the Thames Valley, the Midlands, north-east England and south Wales have very poorly developed lichen floras.

KEEPING RECORDS

By making regular visits at different times of the year and taking samples of vegetation for identification, it should be possible to gradually compile a complete list of species growing in the wood. Information on animal, bird and insect populations present at different times of the year will also be useful. The information can be compared with future surveys to analyse changes to the woodland flora and fauna as a result of the particular management practices employed.

Table A1 *Native Trees (in Approximate Order of Arrival in Britain)*[3]

Common juniper – *Juniperus communis*
Downy birch – *Betula pubescens*
Silver birch – *Betula pendula*
Aspen – *Populus tremula*
Scots pine – *Pinus silvestris*
Bay-leaved willow – *Salix pentandra*
Common sallow – *Salix cinerea* subs *oleifolia*
Common alder – *Alnus glutinosa*
Hazel – *Corylus avellana*

Small-leaved lime – *Tilia cordata*
Bird cherry – *Prunus padus*
Goat willow – *Salix caprea*
Wych elm – *Ulmus glabra*
Rowan – *Sorbus aucuparia*
Almond-leaved willow – *Salix triandra*
Sessile oak – *Quercus petraea*
Ash – *Fraxinus excelsior*
Holly – *Ilex aquifolium*
Alder buckthorn – *Frangula alnus*
Common oak – *Quercus robur*
Elder – *Sambucus nigra*
Hawthorn – *Crataegus monogyna*
Crack willow – *Salix fragilis*
Black poplar – *Populus nigra* var *betulifolia*
Yew – *Taxus baccata*
Blackthorn – *Prunus spinosa*
White beam – *Sorbus aria*
Spindle tree – *Euonymus europaeus*
Midland thorn – *Crataegus laevigata*
Buckthorn – *Rhamnus cathartica*
Crab apple – *Malus sylvestris*
Wild cherry – *Prunus avium*
White willow – *Salix alba*
Field maple – *Acer campestre*
Wild service tree – *Sorbus torminalis*
Large-leaved lime – *Tilia platyphyllos*
Beech – *Fagus sylvatica*
Wayfaring tree – *Viburnum lantana*
Hornbeam – *Carpinus betulus*
Box – *Buxus sempervirens*

Table A2 *Native Shrubs*

Dogwood – *Cornus sanguina*
Guelder rose – *Viburnum opulus*
Privet – *Ligustrum vulgare*
Spurge-laurel – *Daphne laureola*
Gorse – *Ulex europaeus*

Table A3 *Early Introductions*

Almond – *Prunus dulcis*
Elm, English – *Ulmus procera*
Elm, smooth leaved – *Ulmus carpinifolia*
Medlar – *Mespilus germanica*
Myrobalan plum – *Prunus cerasifera*
Peach – *Prunus persica*
Poplar, white – *Populus alba*
Poplar, grey – *Populus canescens*
True service – *Sorbus domestica*
Wild pear – *Pyrus communis*

Table A4 *Some Ancient Woodland Indicator Plants*

Bluebell – *Hyacinthoides non-scripta*
Columbine – *Aquilegia vulgaris*
Common cow wheat – *Melampyrum pratense*
Early dog violet – *Viola riviniana*
Early purple orchid – *Orchis mascula*
Goldilocks – *Ranunculus auricomus*
Green hellebore – *Helloborus viridis*
Herb Paris – *Paris quadrifolia*
Lily of the valley – *Convallaria majalis*
Moschatel – *Adoxa moschatellina*
Nettle-leaved bellflower – *Campanula trachelium*
Opposite-leaved golden saxifrage – *Chrysosplenium oppositifolium*
Polypody – *Polypodium vulgare*
Primrose – *Primula vulgaris*
Ramsons – *Allium ursinum*
Solomon's seal – *Polygonatum multiflorum*
Spurge laurel – *Daphne laureola*
Wild daffodil – *Narcissus pseudonarcissus*
Wood spurge – *Euphorbia amygdaloides*
Wood sorrel – *Oxalis acetosella*
Wood horsetail – *Equisetum sylvaticum*
Wood anemone – *Anemone nemorosa*
Wood sedge – *Carex sylvatica*
Wood rushes – *Luzula spp*
Woodruff – *Galium odoratum*
Yellow pimpernel – *Lysimachia nemorum*
Yellow archangel – *Lamiastrum galeobdolon*

Appendix 2

A Plan of Operations for a Small Wood

Woodland Name: Nutwood

Grid Reference: SP 123456

Local Authority District: West Oxbridge

Special Areas: 3.5 hectares comprise seminatural ancient woodland.

Site Description: Gentle north-facing slope. Access by a grass ride onto public highway. Fences in poor order. Soils are slowly permeable, seasonally waterlogged loams over clay. Wood comprises a mixture of oak, of about 130 years, and ash in the range of 50 to 130 years, with a hazel coppice understorey. Some oak good quality but majority poor. Ash, comprising about 25 per cent of the broadleaved area, is tall and spindly but of good form. There is occasional crab apple, dogwood, spindle and wild service.

Long-Term Aims: To gradually diversify structure of broadleaved area by felling small groups at irregular intervals and restocking with native broadleaves.

Objectives:

- Maintain and create new wildlife habitats.
- Enhance landscape values.
- Produce wood and marketable timber.

Five-Year Work Proposals:

- Repair and maintain fences.
- Fell three informal-shaped coupes, centred upon areas of poorest broadleaved growth, in years one, three and five. Extract timber along route depicted on map. Burn lop and top ensuring no fire within 15 metres of any retained tree. Felling area = 1.5 hectares.
- Replant two of the three felled areas at 2.5 metres x 2.5 metres spacing in 1.2-metre treeshelters, in the first season after felling, using an intimate mixture of: oak (55 per cent); ash (20 per cent); wild cherry (10 per cent); birch (10 per cent); rowan (5 per cent). Recoppice the hazel understorey. Replanting area = 1.0 hectare.
- Allow the third felled area to develop naturally and manage as a wildlife glade. Area = 0.5 hectare.
- Selectively thin the remainder of the broadleaved area removing approximately 25 per cent of the oak (by number) as marked by slash marks. Area = 1.0 hectare.
- Thin the ash to reduce numbers by 30 per cent, leaving a good range of ages. Trees not marked. Area = 1.0 hectare.
- Spot-weed the replanted areas as necessary using approved herbicide.
- Control grey squirrels as necessary using 'trap-door-type' warfarin-baited hoppers.

Appendix 3

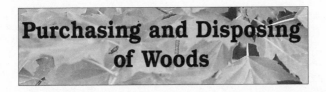

Purchasing and Disposing of Woods

For those who enjoy physical exercise, a sense of achievement, the pride of ownership and the knowledge that they are contributing in some small way to the preservation of Britain's small woodlands, the purchase of a wood must rate serious consideration.

Woodlands that operate as a commercial enterprise are not subject to inheritance tax after a two-year period of ownership; and the sale of timber does not attract tax on income or capital gains, although the underlying land could be subject to an assessment. A number of UK businesses and individuals specialize in forestry and woodland conveyancing and are able to arrange financial, agency and management advice. When purchasing or disposing of a woodland, it is always wise to deal with someone experienced in this particular field. This should ensure that the many pitfalls to be found in the complex business of woodland acquisition and disposal are avoided.[1]

PURCHASE

Prospective purchasers need to consider carefully the following factors before finalizing a purchase.

Access

By far the most common legal problem in woodland ownership is lack of adequate access. Woods are frequently located off the public roads, and even if they do have road frontage it is often difficult to create a suitable access. Does the wood adjoin a public road? If not, does it have the benefit of an adequate right of way to enable general access and timber extraction?

Boundaries

Do the boundaries on the sale plan accord with those on the deed plans? Do the boundaries on the deeds accord with those on the ground?

Fencing

Are there any fencing commitments affecting the wood? Are they still enforceable?

Minerals

Have the minerals under the wood been reserved to a former owner? Are there adequate compensation provisions for the woodland owner if the rights of mineral extraction are exercised?

Timber

Has the timber been reserved to a previous owner?

Sporting

Are the sporting rights included in the sale? What rights does the owner have to control deer? If the sporting rights are included, are they to be let?

Private Rights

Do adjoining owners have rights over the wood (for example, rights of way, rights to use woodland roads, drainage rights)?

Public Rights

Does the public have rights of way over the wood? Are these in the form of footpaths, bridleways or byways?

Commons

Is the wood registered as a common under the Commons Registration Act?

Squatters

Does the woodland have a history of adverse possession by squatters or by encroachment?

Legal Obligations

Is the wood the subject of a Forestry Commission deed of covenant or a conditional felling licence? If grants have been received, who is responsible for their repayment if the crop fails? Is the wood or any of the trees subject to a tree preservation order?

Other Designations

Is the wood in an Area of Outstanding Natural Beauty, an Environmentally Sensitive Area or a Conservation Area? Is the wood, or some part of it, a Site of Special Scientific Interest?

DISPOSAL

Anyone with a small wood to dispose of may wish to discuss prospects with the Woodland Trust (WT). The objectives of the Woodland Trust – a registered charity – are to conserve, restore and reestablish trees and, in particular, broadleaved trees. While it does not concern itself only with small woodlands, acquiring woods is at the heart of the WT's work. Once in their care, woods are open to everyone to enjoy for quiet informal recreation.

Appendix 4

Useful Addresses

ADAS
Agricultural Development and
Advisory Service
Oxford Spires Business Park
The Boulevard
Kidlington
Oxford OX5 1NZ
Tel: 01865 845 122

Arboricultural Association
Ampfield House
Ampfield
Hants SO51 9PA
Tel: 01794 368 717

Association of Professional
Foresters
7–9 West Street
Belford
Northumberland NE70 7QA
Tel: 01668 213 937

British Charcoal Group
72 Woodstock Road
Loxley
Sheffield S6 6TG
Tel: 0114 234 4932

British Christmas Tree Growers
Association
18 Cluny Place
Edinburgh EH10 4RL
Tel: 0131 447 0944

British Deer Society
Burgate Manor
Fordingbridge
Hants SP6 1EF
Tel: 01425 655 434

British Field Sports Society
Old Town Hall
367 Kennington Road
London SE11 4PT
Tel: 0171 582 5432

British Horse Loggers Association
(now merged with the Forestry
Contracting Association)

British Trust for Conservation
Volunteers
36 St Mary Street
Wallingford
Oxford OX10 0EU
Tel: 01491 839 766

CPRE
Council for the Protection of Rural
England
Warwick House
25 Buckingham Palace Road
London SW1W 0PP
Tel: 0171 976 6433

Country Landowners Association
16 Belgrave Square
London SW1X 8PQ
Tel: 0171 235 0511

Countryside Commission England
John Dower House
Crescent Place
Cheltenham
Glos GL50 3RA
Tel: 01242 521 381

Countryside Council for Wales
Plas Penrhos
Fford Penrhos
Bangor
Gwynedd LL57 2LQ
Tel: 01248 385 500

David & Charles (Publishers) Ltd
South Devon House
Newton Abbot
Devon
Tel: 01626 361 1211

English Heritage
23 Savile Row
London W1X 1AB
Tel: 0171 973 3000

English Nature
Northminster House
Peterborough PE1 1UA
Tel: 01733 455 000

Environment Agency
Rivers House
Waterside Drive
Aztec West
Almondsbury
Bristol BS12 4UD
Tel: 01454 624 400

Farming Press
Wharfedale Road
Ipswich IP1 4LG
Tel: 01473 241 122

Farming and Wildlife Advisory
Group
National Agricultural Centre
Stanley
Kenilworth
Warks CV8 2RX
Tel: 01203 696 699

FASTCo
Forestry & Arboricultural Safety &
Training Council
231 Corstorphine Road
Edinburgh EH12 7AT
Tel: 0131 334 8083

Forestry Authority
231 Corstorphine Road
Edinburgh EH12 7AT
Tel: 0131 334 0303

Forestry Authority Research
Station
Alice Holt
Wrecclesham
Farnham
Surrey GU10 4LH
Tel: 01420 22255

Forestry Commission
231 Corstorphine Road
Edinburgh EH12 7AT
Tel: 0131 334 0303

Forestry Contracting Association
Ltd
Dalfling
Blairdaff
Inverurie
Aberdeenshire AB51 5LA
Tel: 01467 651 368

Forestry Trust for Conservation
and Education
The Old Estate Office
Englefield Road
Reading
Berks RG7 5DZ
Tel: 01734 323 523

Game Conservancy
Burgate Manor
Fordingbridge
Hants SP6 1EF
Tel: 01425 652 381

H M Land Registry
Lincoln's Inn Fields
London WC2 3PH
Tel: 0171 917 8888

Health & Safety Commission
Rose Court
2 Southwark Bridge
London SE1 9HS
Tel: 0171 717 6000

Institute of Chartered Foresters
7a St Colme Street
Edinburgh EH3 6AA
Tel: 0131 225 2705

MAFF
Ministry of Agriculture, Fisheries
and Food
55 Whitehall
London SW1A 2EY
Tel: 0171 238 6000

National Farmers Union
164 Shaftesbury Avenue
London WC2H 8HL
Tel: 0171 331 7200

NSWA
National Small Woodlands
Association
3 Perkins Beach Dingle
Stiperstones
Shrops SY5 0PF
Tel: 01743 792 644

Ordnance Survey (OS)
Customer Services
Romsey Road
Maybush
Southampton SO16 4GU
Tel: 01703 792 000

Pesticides Safety Directorate
Mallard House
3 Peasholme Green
Kingspool
York YO1 2PX
Tel: 01904 640 500

Ramblers Association
1/5 Wandsworth Road
London SW8 2XX
Tel: 0171 582 6878

Red Deer Commission
82 Fairfield Road
Inverness IV3 5LH
Tel: 01463 231 751

Royal Forestry Society of England,
Wales & Northern Ireland
102 High Street
Tring
Herts HP23 4AF
Tel: 01442 822 028

RSPB
Royal Society for the Protection of
Birds
The Lodge
Sandy
Beds SG19 2DL
Tel: 01767 680 551

Rural Development Commission
141 Castle Street
Salisbury
Wilts SP1 3TP
Tel: 01722 336255

Scottish Forestry Trust
5 Dublin Street Lane South
Edinburgh EH1 3PX
Tel: 0131 479 7044

Scottish Natural Heritage
12 Hope Terrace
Edinburgh EH9 2AS
Tel: 0131 447 4784

Scottish Timber Trade Association
John Player Building
Stirling Enterprise Park
Springbank Road
Stirling FK7 7RS
Tel: 01786 451 623

Soil Association
Responsible Forestry Programme
& Woodmark Certification
Bristol House
40–56 Victoria Street
Bristol BS1 6BY
Tel: 0117 929 0661

Timber Growers Association
5 Dublin Street Lane South
Edinburgh EH1 3PX
Tel: 0131 588 7111

Tree Council
51 Catherine Place
London SW1E 6DY
Tel: 0171 828 9928

Wildlife Trust
The Green
Witham Park
Waterside South
Lincoln LN5 7JR
Tel: 01522 544 400

Woodland Trust
Autumn Park
Grantham
Lincolnshire NG31 6LL
Tel: 01476 581 111

Appendix 5

Further Reading

Aaron, J R and Richards, E G (1990) *British Woodland Produce*, Stobart Davies, London.

Anderson, M L (1950) *The Selection of Tree Species*, Oliver & Boyd, Edinburgh & London (2nd edition, 1961).

Anon (1988) *Farm Conservation Guide*, Schering Agriculture, Nottingham.

Anon (1981) *Field Guide to the Trees and Shrubs of Britain*, The Readers Digest Association Ltd, New York (Reprinted 1985).

Baines, C (1985) *How to Make a Wildlife Garden*, Hamish Hamilton, London.

Bevan, D (1987) 'Forest Insects' *Forestry Commission Handbook* 1, HMSO, London.

Blythe, J, Evans, J, Mutch, W E S and Sidwell, C (1987) *Farm Woodland Management*, Farming Press, Ipswich (2nd edition, 1991).

Boulton, E H B and Jay, B A (1944) *British Timbers – Their Properties, Uses and Identification*, Adam and Charles Black, London.

Bowles J (ed) (1995) *Small Woods, Who Cares?* Conference Proceedings, Blenheim Palace, 17 May 1995. Oxfordshire Woodland Project and National Small Woodland Association.

Brazier, J D (1990) 'The Timbers of Farm Woodland Trees' *Forestry Commission Bulletin* 91, HMSO, London.

Broad, K (1989) 'Lichens in Southern Woodlands' *Forestry Commission Handbook* 4, HMSO, London.

Brown, R G (1972) 'Recreation Demand and Countryside Policies for Meeting It' in Timber Growers Organization *Recreation and the Woodland Owner*, Beaconsfield Conference Proceedings 1972.

Buckley, G P (ed) (1992) *Ecology and Management of Coppice Woodlands*, Chapman & Hall, London.

Countryside Commission, Country Landowners Association and the National Farmers Union (1994) *Managing Public Access: A Guide for Farmers and Landowners*, Countryside Commission, Cheltenham.

Countryside Recreation Network (1996) *The 1994 UK Day Visits Survey*, Countryside Recreation Network, Cardiff.

Crowther, R E and Evans, J (1984) 'Coppice' *Forestry Commission Leaflet* 83, HMSO, London.

Evans, J (1988) 'Natural Regeneration of Broadleaves' *Forestry Commission Bulletin* 78, HMSO, London.

Evans, J (1984) 'Silviculture of Broadleaved Woodland' *Forestry Commission Bulletin* 62, HMSO, London.

Ferris-Kaan, R (ed) (1995) *The Ecology of Woodland Creation*, John Wiley & Sons Ltd, Chichester.

Forestry Commission and Countryside Commission for Scotland (1984) *Providing for Children's Play in the Countryside*, Countryside Commission for Scotland, Perth.

Forestry Trust for Conservation and Education (1995) *Woodlands to Visit in England and Wales*, Forestry Trust, Reading.

Garfitt, J E (1995) *Natural Management of Woods: Continuous Cover Forestry*, Research Studies Press, Taunton, Somerset.

Grayson, A J (1993) *Private Forestry Policy in Western Europe*, CAB International, Wallingford, Oxon.

Greig, D A (ed) (1993) *Farm and Small Scale Forestry*, Proceedings of a Discussion Meeting, University of Reading, 3–5 April 1992, Institute of Chartered Foresters, Edinburgh.

Hampshire County Council (1995) *Hazel Coppice: Past, Present and Future*, Hampshire.

Harding, T (1988) 'British Softwoods Properties and Uses' *Forestry Commission Bulletin* 77, HMSO, London.

Harris, E & Harris, J (1991) *Wildlife Conservation in Managed Woodlands & Forests*, Basil Blackwell, Oxford.

Hart, C (1991) *Practical Forestry for the Agent and Surveyor*, Alan Sutton Publishing Ltd, Gloucester.

Hibberd, B G (ed) (1991) 'Forestry Practice' *Forestry Commission Handbook* 6, HMSO, London.

Hibberd, B G (1988) 'Farm Woodland Practice' *Forestry Commission Handbook* 3, HMSO, London.

Hine, A (ed) (1995) *Woodland Pond Management*, Proceedings of Corporation of London meeting, 7 July 1994, Richmond Publishing Co Ltd.

Hodge, S J (1995) 'Creating and Managing Woodlands Around Towns' *Forestry Commission Handbook* 11, HMSO, London.

Holmes, F (1974) *Following the Roe*, John Bartholomew & Son Ltd, Edinburgh.

Hudson, D (1995) *The Roughshooter's Handbook*, Swan Hill Press, Shrewsbury.

Insley, H (ed) (1988) 'Farm Woodland Planning' *Forestry Commission Bulletin* 80, HMSO, London.

James, N D G (1991) *Forestry & Woodland Terms*, Blackwell, Oxford.

James, N D G (1989) *The Forester's Companion*, fourth edition, Blackwell, Oxford.

Kelly, D W (1986) *Charcoal and Charcoal Burning*, Shire Album 159, C I Thomas & Sons (Haverfordwest) Ltd.

Kerr, G and Evans, J (1993) 'Growing Broadleaves for Timber' *Forestry Commission Handbook* 9, HMSO, London.

Kirby, K J (1991) *Regional Patterns and Woodland Management in British Woods*, Nature Conservancy Council, CSD Note 56, Peterborough.

Lack, P (1992) *Birds on Lowland Farms*, HMSO, London.

Leathart, S (1991) *Whence our Trees?*, Foulsham, Berks.

Lorrain-Smith, R (1996) *Grants for Trees*, Calderdale Metropolitan Borough Council, West Yorkshire.

Maclean, M (1992) *New Hedges for the Countryside*, Farming Press, Ipswich.

MAFF (1993) *Farm Woodlands: A Practical Guide*, HMSO, London.

MAFF/Health and Safety Commission, (1990) *Pesticides: Code of Practice for the Safe Use of Pesticides on Farms and Holdings*, HMSO, London.

Marren, P (1992) *The Wild Woods: A Regional Guide to Britain's Ancient Woodland*, David & Charles, Newton Abbot.

Marren, P (1990) *Britain's Ancient Woodland: Woodland Heritage*, David & Charles, Newton Abbot.

Mattheck, C and Breloer, H (1994) 'The Body Language of Trees' *Research for Amenity Trees* 4, Department of the Environment, HMSO, London.

McCall, I (1988) *Woodlands for Pheasants*, The Game Conservancy, Hampshire.

Milner, J E (1992) *The Tree Book*, Collins & Brown, London.

Mitchell, A (1974) *A Field Guide to the Trees of Britain and Northern Europe*, Collins, London (reprinted 1986).

Pepper, H W (1978) 'Chemical Repellents' *Forestry Commission Leaflet* 73, HMSO, London.

Porter, V (1990) *Small Woods and Hedgerows*, Butler & Tanner Ltd, Frome and London.

Potter, M J (1991) 'Treeshelters' *Forestry Commission Handbook* 7, HMSO, London.

Prior, R (1987) *Deer Management in Small Woodlands*, The Game Conservancy, Fordingbridge, Hants.

Probert C (1989) *Pearls in the Landscape: The Conservation and Management of Ponds*, Farming Press Books, Ipswich.

Rackham, O (1990) *Trees and Woodland in the British Landscape*, Dent, London.

Rackham, O (1986) *The History of the Countryside*, Weidenfeld & Nicolson, London.

Rackham, O (1980) *Ancient Woodland: Its History, Vegetation and Uses in England*, Edward Arnold, London.

Ratcliffe, P R and Mayle, B A (1992) 'Roe Deer Biology and Management' *Forestry Commission Bulletin* 105, HMSO, London.

Shaw, G and Dowell, A (1990) 'Barn Owl Conservation in Forests' *Forestry Commission Bulletin* 90, HMSO, London.

Robertson, P A (1992) 'Woodland Management for Pheasants' *Forestry Commission Bulletin* 106, HMSO, London.

Rodwell, J and Patterson, G (1994) 'Creating New Native Woodlands' *Forestry Commission Bulletin* 112, HMSO, London.

Savill, P S (1991) *The Silviculture of Trees used in British Forestry*, CAB International, Wallingford, Oxon.

Sparkes, I G (1977) 'Woodland Craftsmen' *Shire Album* 25, C I Thomas & Sons (Haverfordwest) Ltd (reprinted 1991).

Strouts, R G and Winter, T G (1994) 'Diagnosis of Ill-health in Trees' *Research for Amenity Trees* 2, Forestry Commission, Edinburgh.

Tabor, R (1994) *Traditional Woodland Crafts*, B T Batsford Ltd, London.

Thomas, G S (1983) *Trees in the Landscape*, Jonathan Cape Ltd, London.

Walker, W (1993) *A New Strategy for the Oxfordshire Woodland Project*, Canopy Consultants, Shropshire.

Watkins, C (1990) *Woodland Management and Conservation*, David & Charles, Newton Abbot.

White, J (1995) *Forest and Woodland Trees in Britain*, Oxford University Press, Oxford.

Wilkinson, G (1978) *Epitaph for the Elm*, Hutchinson & Co, (Publishers) Ltd, London.

Williamson, D R (1992) 'Establishing Farm Woodlands' *Forestry Commission Handbook* 8, HMSO, London.

Willoughby, I and Clay, D (1996) 'Herbicides for Farm Woodlands and Short Rotation Coppice' *Forestry Commission Field Book* 14, HMSO, London.

Willoughby, I and Dewar, J (1995) 'The Use of Herbicides in the Forest' *Forestry Commission Field Book* 8, HMSO, London.

OTHER FORESTRY COMMISSION PUBLICATIONS

'Classification and Presentation of Softwood Sawlogs' *Forestry Commission Field Book* 9

'Forest Mensuration Handbook' *Forestry Commission Booklet* 39

Forest Nature Conservation Guidelines

Forest Recreation Guidelines

Forestry and Archaeology Guidelines

Forests & Water Guidelines

Lowland Landscape Design Guidelines

The Management of Semi-natural Woodlands

Wildlife Rangers Handbook

Glossary of Woodland Terms

Afforestation: The planting of trees on a formerly unwooded area.

Agroforestry: That branch of forestry concerned with the growing of trees on land also used for the production of crops or livestock.

Ancient woodland: A site that has been wooded continuously since at least AD1600.

Annual coupe: The total area of felling in a year.

Annual ring: A layer of wood, representing one year's growth, seen in the transverse section of the stem of a tree.

Arable coppice: see **Short rotation coppice**.

Arboretum: A place where trees or shrubs are cultivated for their scientific or educational interest.

Arboriculture: The cultivation of trees and shrubs to produce specimens primarily for ornament, landscape, shelter and other objectives other than for timber production.

Assart: An historical term for an area converted into arable land by grubbing up trees and other growth from forest land.

Beam: A squared section of timber suitable for constructional use, typically with a minimum length of 6 metres (20 ft) and minimum sides of 20 centimetres (eight inches) and cut from the heartwood of a straight oak log with no large side branches.

Beat up: To replace failed trees within the first few years of planting.

Belt: A narrow strip of trees.

Besom: A yard brush made of bundled twigs, typically of birch or heather.

Biomass: The exploitation of plant crops that may be combusted for their heat energy or processed into fuels. Biomass represents a renewable source of energy.

Blaze: To mark a tree by removing a slice of bark from the main stem – usually to identify it for felling.

Bole: The main stem of a tree up to its first major branch.

Bolling: The permanent main stem of a pollarded tree.

Brake: An area of dense undergrowth, shrubs, brushwood, etc.

Brash: (a) The small branches cut off trees. (b) To prune branches from young trees up to about two metres (six feet) above ground level.

Broadleaf: Deciduous or evergreen dicotyledon trees.

Burr: An excrescence of rough growth that may develop on the stem of a tree. Such stems may be greatly valued for veneers.

Butt: The log section nearest the root.

Cambium layer: The thin layer of growing cells beneath the bark of a tree.

Canker: An open wound on a tree or shrub, usually caused by a fungus or bacteria.

Canopy: The topmost layer of a woodland consisting of the foliage and branches growing together above head height.

Cant: A section of coppice harvested on a regular basis.

Carr: Wet, boggy woodland usually dominated by alder.

Chase: An unenclosed area of land where wild animals were preserved to be hunted. Unlike a forest, a chase could be held by a subject of the Crown.

Cleaning: The removal of unwanted trees and shrubs that have usually arisen from natural regeneration or coppice regrowth from within a plantation.

Clearfelling: Felling of a whole woodland or compartment of trees at one time.

Clinometer: An instument used for measuring angles of inclination.

Clone: A number of genetically identical trees or plants.

Clump: A small group of trees.

Commercial timber production: Management to produce the maximum financial return from tree crops.

Compartment: A management area within a woodland that is given an individual name or number.

Conifer: Deciduous or evergreen gymnosperm trees with needle-like or scale-like leaves, generally bearing woody cones.

Coppice: (a) An area of broadleaved trees, traditionally grown for rotations of eight to 25 years, at the end of which time the wood is cut near ground level and then allowed to reproduce itself by means of shoots thrown up from the stumps or stools. (b) The multistemmed underwood trees or shrubs created by coppicing. (c) To manage trees by cutting the stems at intervals for the production of smallwood products.

Coppice cycle: The number of years between cutting coppice.

Coppice-with-standards: A traditional method of management that combines the growing of trees at relatively wide spacing for timber production, with an understorey of coppice trees or shrubs growing between them.

Copse: A woodland managed for producing coppice; another name for a coppice.

Cord: (a) A stack of wood, usually eight feet by four feet by four feet, cut from branches of trees and typically used for firewood or charcoal production. (b) To cut wood in appropriate lengths and stack in a cord.

Cordwood: Branches suitable for cutting into lengths equal to the width of a cord.

Coupe: A clearfell section in a coppice or woodland.

Covert: A thicket or woodland providing cover for game.

Crown: The upper part of a tree that includes the branches.

Crown thinning: A type of thinning where trees are removed with the express purpose of relieving crown competition of selected retained trees.

Cube: See **Hoppus foot.**

Cull: (a) A weak or misshapen seedling or transplant that is usually discarded. (b) To discard such seedlings or transplants.

Dapple shade: A method of management designed to retain a scatter of trees in a woodland, typically between one third and one half of the usual full stocking density.

Deciduous: Shedding all the leaves periodically; the opposite of evergreen.

Deer glade: A grassy area within a woodland, used for deer population management.

Deer lawn: See Deer glade.

Dicotyledon: A flowering plant having two embryonic seed leaves.

Easement: The right enjoyed by a landowner to make limited use of neighbouring land – for example, by crossing it to reach his or her property.

Eclectic thinning: A type of thinning where a proportion of all sizes, species, ages and quality types are removed.

Enrichment: Adding trees in gaps in an established woodland or plantation by planting or natural regeneration.

Epicormic shoots: Twigs sprouting from dormant or adventitious buds on a tree's main stem.

Establishment stage: The period in the life of a young plantation from the time of planting to the stage when no further beating-up or weeding is required.

Evergreen: Retaining green leaves throughout the year; the opposite of deciduous.

Extraction: The removal of felled timber from a woodland.

Extraction rack: A track used for the extraction of timber, typically created by cutting out one or more rows of trees at the time of first thinning.

Faggot: A bundle of small branches bound together for use as a fuel for ovens.

Feathered tree: A young nursery tree of about two metres, well furnished with side branches and with a straight upright leader, suitable for planting.

Fencing bar: A log suitable for converting into fencing panel material, typically of softwood with a minimum length of 1.85 metres (6.1 feet) and a minimum top diameter of 16 centimetres (six inches).

Forest: Historically, a prescribed but unenclosed area of land on which the sovereign had the exclusive right to hunt and make forest laws. More commonly, a large wooded area.

Forest transplant: A nursery tree with a sturdy stem and good fibrous rootstock, usually between 12 centimetres and 50 centimetres tall, suitable for planting.

Forest year: The forest year runs from 1 October to 30 September and is recorded as the second of the two calendar years involved (for example, an event in November 1996 is recorded as Forest Year 1997, or FY 97).

Forestry: The practice of growing trees in plantations.

Formative pruning: The removal of branches, typically within three to 10 years of planting, with the objective of improving timber quality.

Full stocking: A standard measure of tree quantity and distribution of various species in a woodland accepted as a full complement, typically for the payment of grant aid.

Girdling: See **Ring barking.**

Glade: An area of herbaceous growth or an unplanted zone within a forest or woodland.

Good form: Trees with relatively straight stems and with branches of relatively small diameter.

Greenwood: Freshly felled wood.

Grove: A forestry plantation.

Group felling: Felling of a group of trees or a subcompartment within a woodland.

Hanger: A wood growing on the side of a steep hill.

Harden off: The process by which plants become resistant to cold, frost, etc, by gradual exposure to such conditions.

Hardwood: The timber of broadleaved trees.

Heartwood: The inner core of a tree stem that no longer conducts sap.

Hectare: A metric land measure containing 100 acres. One hectare (ha) = 10,000 square metres. One hectare = 2.471 acres.

High forest: Woodland or forest in which the majority of the trees are being grown to full size with a closed canopy.

High pruning: The removal of branches, typically of selected trees, above the height normally achieved by brashing to improve timber quality.

Holt: A wooded hill.

Hoppus foot: An imperial measure of log volume based on length and quarter-girth at mid point. A Hoppus foot is about 21 per cent short of a true cubic foot, the reduction allowing for loss of volume (from sawdust and other waste) when converting the round log to sawn wood.

■ 1 Hoppus foot = 1.27 true cubic feet.
■ 1 Hoppus foot = 0.036 cubic metres.
■ 1 cubic metre = 27.74 Hoppus feet.

Layering: Fixing a living stem of a tree or shrub into the ground to enable roots to develop at or near the fixed point, for the purpose of growing another tree or shrub. Often used to create new coppice plants.

Leader: The single topmost shoot of a tree representing the most recent season's growth.

Leading shoot: See **Leader.**

Lime-induced chlorosis: Impaired chlorophyll synthesis resulting in yellowing foliage and die-back.

Line thinning: A type of thinning where complete rows of trees are removed at regular intervals, typically as part of a first thinning operation, to maximize output, to minimize cost and facilitate timber extraction.

Log: Any round wood. See **Sawlog.**

Logging arch: A wheeled device used to transport logs with the butt ends held clear of the ground.

Lop and top: The branches and other unused parts cut from a felled tree.

Lower canopy: The canopy of the understorey.

Maiden tree: Any tree not grown from a coppiced stump.

Mast year: Any year in which trees produce exceptionally large quantities of seed and fruit; typically occurs at five- to seven-year intervals.

Monoculture: Where reliance is placed on just one species as a tree crop.

Mulch mat: A mat of some light-inhibiting material, typically black polythene, capable of suppressing or preventing weeds from germinating around the bases of newly planted trees.

Native species: A species that was not originally introduced by human agency.

Natural regeneration: Trees and shrubs grown in situ from the naturally shed seed of adjacent trees or shrubs.

Netlon guard: A type of guard comprised of polyethylene mesh used to protect young trees.

Nurse species: Hardy, quick-growing trees planted and managed for the express purpose of sheltering less vigorous, less robust or more valuable tree species when they are young.

Open ground: An area of herbaceous growth within a woodland, sometimes containing occasional trees and shrubs.

Overstorey: Trees that make up the upper canopy of a woodland.

Panel: A section of coppice cut in one year.

Permissive right of way: A path or ride, other than a right of way, for use by the general public by agreement of the landowner.

Photosynthesis: The synthesis of organic compounds from carbon dioxide and water using light energy absorbed by chlorophyll.

Pinetum: An area of land where pine trees and other conifers are grown.

Pioneer species: A tree or shrub species capable of first colonizing open ground before any other slower-growing species can become established.

Plantation: An area of closely planted trees other than an orchard.

Planting year: The year in which a planted tree crop first grew. Trees planted in autumn 1996 or spring 1997 (the dormant season) would both be recorded as P 97. See **Forest year**.

Ploughpan: A hard impervious layer found in some soils caused by ploughing to the same depth over a long period.

Podsol: A type of soil having a greyish-white colour in its upper layers from which certain minerals have been leached.

Pole: A young tree or coppice growth.

Pole stage: For conifers, trees from first thinning to 40 years. For broadleaves, trees from first thinning to 50 years.

Poll: Another name for pollard.

Pollard: A tree formed by pollarding.

Pollarding: A traditional form of management akin to coppicing but in which the tree is cut off above the browsing height of livestock – typically between 1.5 metres and 3.5 metres above ground level – and allowed to regrow to produce a multiple stemmed crown. Recutting takes place on a regular cycle.

Polling: Another name for pollarding.

Pressler borer: A hand-held drilling device used to extract a narrow core of wood from the stems of trees.

Pruning: The removal of the branches of a tree or shrub, typically to about two metres (six feet) above ground level, to remove dangerous branches or to improve timber quality.

Pulpwood: Low-quality roundwood of a size suitable for the production of paper or board.

Quarter-girth: A traditional measure used in estimating the volume of a felled log. It is based on the quarter-girth inch which is four true inches. See **Hoppus foot.**

Rack: See **Extraction rack.**

Regeneration: Management practised to replace trees in a woodland by replanting (artificial regeneration) or by natural regeneration.

Relascope: Any of a number of hand-held devices used to estimate the cross-sectional area of timber in a crop from a measure of tree diameters.

Replanted ancient woodland: An ancient woodland site that has been replanted.

Replanting: Planting an area cleared of trees or shrubs.

Respace: To cut out surplus young trees before they have reached the thinning stage. Refers in particular to naturally regenerated trees.

Ride: A permanent unsurfaced route within a woodland, used for access, demarcation, extraction of timber and shooting purposes.

Ring-barking: The removal of a tree's bark and cambium in a complete ring around the stem. Such action will result in the tree's death.

Ring shake: Where the wood splits along the annual rings.

Rotation: The number of years from planting to felling age of a tree crop.

Roundwood: Timber of small diameter that has not been converted. Typically for use as stakes, rustic poles, etc.

Sapwood: The living wood, between the dead cells of the heartwood and the cambium layer beneath the bark, that conducts water from the soil up the tree.

Sawlog: Timber of a size and quality acceptable to a sawmill, typically straight, clean stems at least 16 centimetres (six inches) in diameter and at least three metres (10 feet) long.

Scalloping: The removal of scallop-shaped sections of growth along the edges of woodland rides and boundaries to enhance landscape or nature conservation values or to improve sporting potential.

Scarify: To disturb the soil surface, typically to encourage germination of tree seed.

Screen: A wood that serves to shelter, protect or conceal.

Scrub: An area of low growth and bushes.

Secondary woodland: Woodland growing on a site that was formerly not woodland.

Seed trees: Trees grown to produce seed for the production of tree seedlings either in situ or in a tree nursery.

Selective felling: Felling to remove only selected trees of particular commercial value.

Selective thinning: A type of thinning where trees are removed to give the best-quality trees, or preferred species, more room to grow.

Seminatural woodland: A woodland site with trees of mainly native species that are of natural occurrence (not originated from planting).

Seminatural ancient woodland: An ancient woodland site with trees of mainly native species which are of natural occurrence (not originated from planting).

Shake: A split defect in felled timber.

Shaw: A narrow strip of woodland growing along the margin of a field.

Shelterbelt: A strip of trees planted to provide shelter from winds.

Shoot: Any new growth of a plant.

Short rotation coppice: A modern management method using fast-growing trees, typically poplar and willow species, planted at relatively close spacing to be cut on a three- to five-year rotation for the production of wood chips.

Shrub: A woody plant whose normal habit is to grow without a single main stem.

Shrub layer: That part of the woodland structure containing shrubs and young tree growth.

Silvicultural thinning: A type of thinning designed to maximize wood production.

Singling: Reducing the number of stems on a coppiced stool to leave just one on which to grow.

Slabbing: Semiround sawmill offcuts.

Snedding: The removal, or trimming, of branches and top from a felled tree.

Softwood: The timber of coniferous trees.

Spat: Another name for a mulch mat.

Spinney: A small wood, copse or thicket.

Spot weeding: Removing weeds from the immediate vicinity of individual young trees rather than from complete rows.

Spring: An area of young coppice.

Stag-headed: A tree, typically an oak, in which many of the crown's branches have died back and protrude, bare of bark and leaves.

Stand: A group or area of trees.

Standard: (a) A single-stemmed tree that is grown to maturity, typically planted among coppice. (b) A young nursery tree with about two metres (six feet) of clear stem, a well-balanced head of branches and an upright central leader, suitable for planting.

Standing sale: The sale of trees standing, as compared to sale at stump or at roadside.

Star shake: Where the wood splits along the medullary rays.

Stem: The main trunk of a tree.

Stocking density: The number of plants growing per unit area.

Stocking level: See **Stocking density.**

Stool: The stump of a coppiced tree or shrub remaining in the ground after cutting and from which new shoots grow.

Stored coppice: Singled coppice stems retained beyond their normal coppice cycle.

Subcompartment: A subdivision of a compartment, typically comprising tree crops of the same age and species.

Sucker: A shoot arising from an underground root, near to or at a distance from the main stem.

Tally counter: A hand-held device used to keep score, typically when counting trees or measuring distances.

Thicket: A dense growth of small trees, shrubs and similar plants.

Thicket stage: The period when the branches of a tree crop meet and become intermingled.

Thinning: The removal of a proportion of trees in a crop to afford more growing space to the retained trees, or to obtain a supply of timber.

Timber: The stems of trees of large enough diameter to be suitable for producing beams and planks.

Timber gouge: A metal instrument used to make marks in tree bark.

Timber sword: A pliable strip of metal ending in a hook, used to pull a measuring tape beneath large logs.

Tree: A woody plant with a single main stem in natural growth.

Treeshelter: A translucent plastic tube for newly planted trees that guards against browsing predators and encourages the tree's development.

Turnery wood: Wood suitable for producing articles with the aid of a lathe.

Undercut seedling: A nursery plant whose root has been deliberately shortened to encourage the development of a fibrous root system.

Underplanting: The planting of a new crop under an existing one to increase site productivity.

Understorey: Trees and shrubs growing under the canopy of other trees.

Underwood: Coppice woodland, growing or cut.

Upper canopy: The canopy of the dominant trees or shrubs.

Veneer wood: Straight tree stems of sizes suitable for the production of wood veneers by peeling.

Walk: A division of a forest in order to facilitate its administration.

Wattle hurdle: A fencing panel of woven rods, typically of hazel.

Wayleave: Access to a property granted by a landowner for payment – for example, to allow a contractor access to a felling site.

Whip: A young, single-stemmed, nursery tree between about one and two metres, suitable for planting.

Whorl: A radial arrangement of branches around a stem.

Wind snap: Snapping of tree stems by the wind.

Windblow: Trees uprooted either partially or wholly by the wind.

Windrow: A long, low line of branches or other material.

Windthrow: See **Windblow.**

Wolf: A vigorous, coarse, often poorly formed tree of low timber value.

Wood: (a) The poles and branches of trees of a smaller diameter than timber. (b) Another name for woodland.

Wood-pasture: An area of open-grown trees on which farm animals or deer are raised.

Woodbank: A boundary bank surrounding or subdividing a woodland.

Woodland: Land that is mostly covered with dense growths of trees and shrubs; a generalized term.

Woodland type: A description of the overall appearance of a woodland (for example, coppice-with-standards), or a description based on the constituent parts of a woodland (for example, ash/oak acid woodland), or a description based on the apparent age of the woodland (for example, ancient woodland).

Yield class: A system of assessing the productivity of a crop of trees based upon the measurement of tree height and age.

References

INTRODUCTION

1   A Scott (1992) 'Small Woodlands in Britain – A National Perspective' in D A Greig (ed) *Proceedings of a Discussion Meeting of the Institute of Chartered Foresters*, University of Reading, 3–5 April 1992.

2   Pers comm R Melville, Forestry Authority Conservator.

3   Pleydell, G (1993) in John Clegg & Co report (1994) *Markets for Hardwood Timber*, John Clegg & Co, Edinburgh.

4   John Clegg & Co (1994 *The Rural Property Market: The Influences of Woodlands on Property Values in 1993*, John Clegg & Co, Edinburgh.

5   Various references, including:

(a) 'a high proportion of the woods are in an unmanaged and often derelict condition' in 'Forest of Mercia Project Report', *Heartwood* 16, NSWA newsletter, Summer 1996.

(b) 'Detailed research in Norfolk and Suffolk identified that 78 per cent of their woods were in a state of neglect... it can be reasonably assumed that the levels of neglect found in Norfolk and Suffolk also occur in Essex and Cambridgeshire' in 'Anglian Woodland Project Report' by Vincent Thurkettle, *Heartwood* 5, NSWA newsletter, Winter 1992.

(c) 'Many of these small woods are poorly managed or not managed at all' in Oxfordshire's 'Woodland Project Report' by Ken Broad, *Heartwood* 10, NSWA newsletter, Summer 1994.

(d) 'Much of the estimated 35,000 hectares of birchwoods across the Highlands now consists of neglected stands of overmature and moribund trees' in 'Highland Birchwoods Report' by Andrew Thompson, *Heartwood* 9, NSWA newsletter, Spring 1994.

(e) 'To examine the reasons for the decline and demise of upland woodland in the South Pennines', in 'South Pennines Woodland Project Report' by Edward Mills, *Heartwood* 5, NSWA newsletter, Winter 1992.

6   Cate, J (1994) 'Small Woodlands and their Owners: The Silvanus Experience', *Quarterly Journal of Forestry*, April 1994, pp 128–133.

7   Countryside Commission (1983) *Small Woods on Farms*, Countryside Commission, Cheltenham.

CHAPTER 1

1   Forestry Authority (1998) *The UK Forestry Standard*, Forestry Authority, Edinburgh.

2   Harris, E and Harris, J (1991) *Wildlife Conservation in Managed Woodlands and Forests*, Blackwell, Oxford.

3   Forestry Authority (1995) *Forestry and Archaeology Guidelines*, Forestry Authority, Edinburgh.

4   Bannister, N R (1996) *Woodland Archaeology in Surrey*, Surrey County Council, Surrey.

5   Countryside Commission (1990) *The Rights of Way Act 1990 – Guidance notes for farmers*, Countryside Commission, Cheltenham.

6   DoE (1994) *Tree Preservation Orders, A Guide to the Law and Good Practice*. Department of the Environment, London.

7   Ibid.

8   Lamb, A (1995) *Attitudes of Owners to Woodlands in England*. Msc Degree thesis. Oxford Forestry Institute, University of Oxford, Oxford.

9   Gordon, W A (1965) 'Sustained Yield and the Law', *Quarterly Journal of Forestry* 59, pp 52–55.

10 Heritage, S (1996) *Protecting Plants from Damage by the Large Pine Weevil and Black Pine Beetles*, Forestry Commission Research Information Note 268, Farnham.

11 (1996) *Grazing in Upland Woods: Managing the Impacts*. English Nature booklet, Peterborough.

12 Ordnance Survey (1997) *Copyright 1: Copying for Business Use; Copyright 2: Publishing; Copyright 3: Digital Map Data*, Ordnance Survey, Southampton.

13 Anon (1997) 'Personal Protection Equipment: What Should You Be Wearing?' in *Quarterly Journal of Forestry* vol 91, no 1, January, pp 49–50.

14 Forestry Authority (1992) *Rest Allowances and Protective Clothing*. Forestry Authority Technical Development Branch Report 7/92 Dumfries.

15 (1994) *Hand-arm Vibration*. Health & Safety Executive Book HS (G) 88, London.

16 Jackson, J E (1962) 'Lyme Disease', *Quarterly Journal of Forestry*, vol 86, no 2, April 1992, p 113.

17 Plimsoll Publishing Ltd (1997) *Plimsoll Portfolio Analysis – Timber Merchants*. Second edition. Plimsoll Publishing Ltd, Cleveland.

CHAPTER 2

1   Forestry Commission (1981) 'Yield Models for Forest Management', *Forestry Commission Booklet* 48; Forestry Commission (1975) 'Forest Mensuration Handbook', *Forestry Commission Booklet* 39, HMSO, London.

2   Harmer, R (1992) *Do Dominant Oaks have few Epicormic Branches?* Forestry Authority Research Information Note 223, Farnham.

3   Henman, G S and Denne, M P (1992) *Shake in Oak*, Forestry Authority, Research Information Note 218, Farnham.

4   Forestry Commission (1995) *The Adaption of Agricultural Tractors for Forestry*, Forestry Commission, Technical Development Branch, Technical Note 24/95, Dumfries.

5   Forestry Commission (1995) *Timber Trailers for Agricultural Tractors*, Forestry Commission Research Division, Technical Development Branch, Technical Note 28/95, Dumfries.

6   Forestry Commission (1994) *Horse Extraction of Large Hardwood Logs*, Forestry Commission, Technical Development Branch, Report 12/94, Dumfries.

7   Forestry Commission (1993) *Horse Extraction in Thinning*, Forestry Authority, Technical Development Branch, Report 13/93, Dumfries.

8   (1993) *Extraction by the 'Iron Horse' in Broadleaved Woodland*, Forestry Authority Wales, Technical Development Branch, Report 1/93, Dumfries.

9   (1996) *Adding Value to Farm Wood*, Forestry Authority, Research Division Technical Development Branch, Technical Note 21/96, Dumfries.

10 Andrews, J (1995) 'Kilning Made Simple for All Users' *Forestry & British Timber*, July 1995, pp 25–27.

11 North England Liaison Group (1995) *Selling Your Trees*, Forestry Commission, British Timber Merchants Association and Timber Growers Association.

12 Walker, W B (1993) *A New Strategy for the Oxfordshire Woodland Project*, Canopy Consultants, Shropshire.

13 Peterken, G F (1991) 'Managing Semi-natural Woods: A Suitable Case for Coppice', *Quarterly Journal of Forestry*, vol 85, no 1, January 1991, pp 21–29.

14 Nisbet, J (1905) *The Forester*, Blackwood & Sons, Edinburgh.

15 Mitchell, P L (1989) 'Repollarding Large Neglected Pollards: A Review of Current Practice and Results', *Arboricultural Journal*, AB Academic Publishers 1989, vol 13, pp 125–142.

16 Rackham, O (1976) *Trees and Woodlands in the British Landscape*, Dent, London.

17 Read, H J (ed) (1991) *Pollard and Veteran Tree Management*, Conference Proceedings, Corporation of London.

18 Read, H J, Frater, M and Turney, I S (1991) *The Management of Ancient Beech Pollards in Wood Pastures*, Arboricultural Research Note 95/91, DOE Arboricultural Advisory & Information Service, Farnham.

19 Pfetscher, G and Denne, M P (1995) 'Survival and Growth of Repollarded Old Beeches, *Quarterly Journal of Forestry*, vol 89, no 1, January, pp 40–45.

20 Brooks, A (1980) *Woodlands: A Practical Conservation Handbook*, British Trust for Conservation Volunteers, Reading.

21 (1996) *Adding Value to Farm Wood*, Forestry Authority Research Division Technical Development Branch, Technical Note 21/96, Dumfries.

22 Houldershaw, D C (1991) 'Woodchippers Wider Woodland Potential' *Forestry & British Timber*, June 1991, p38.

23 Kelley, D (1995) 'Charcoal – A Source of Secondary Income for the Woodland Owner', *Farming & Conservation*, October 1995, pp 5–7.

24 Desai P (1997) 'Biregional Charcoal Company', speaking at 'The Hazel Revolution' Conference, West Dean College, 24 March 1997.

25 (1995) *Charcoal Production: A 2 Tier Kiln Case Study*, Forestry Commission Research Division, Technical Development Branch, Report 4/95, Dumfries.

26 Snell, H (1996) 'A Second Chance for Poplar', *Woodland Heritage News* pp 4–5.

27 Tabbush, P (1995) *Approved Poplar Clones*, Forestry Authority Research Information Note 265, Farnham.

28 Hunt, J (1996) 'Realising poplar potential', *Farmer's Weekly*, 20 September 1996, pp 75–77.

29 Tabbush, P (1993) 'Poplar Husbandry', *Quarterly Journal of Forestry*, July 1993, pp 203–206.

30 Forestry Commission (1958) *Cultivation of the Cricket Bat Willow*, Bulletin No 17, HMSO, London.

CHAPTER 3

1 Harmer, R and Kerr, G (1995) *Natural Regeneration of Broadleaved Trees*, Forestry Authority Research Information Note 275, Farnham.

2 Harmer, R (1994) 'Selecting Sites for Successful Natural Regeneration of Broadleaves', *Forestry & British Timber*, December 1994, pp 16–18.

3 Tabbush, P M (1986) *Rough Handling Reduces the Viability of Planting Stock*, Arboricultural Research Information Note 64/86, DoE, Arboricultural Advisory and Information Service.

4 Patch, D (1987) *Tree Staking*, Arboricultural Research Note 40/87, DoE, Arboricultural Advisory and Information Service.

5 (1994) *Mulching Trial: Evaluation of Currently Available Mulching Systems*, Forestry Commission, Technical Development Branch Report 3/94, Dumfries.

6 Pepper, H W Chadwick, A H and Buff, R (1992) *Electric Fencing against Deer*, Forestry Commission Research Information Note 206.

7 (1995) *Temporary Deer Fencing*, Forestry Authority, Technical Development Branch, Technical Note 5/95, Dumfries.

8 Hodge, S and Kerr, G (1991) 'Young Trees Need to be Freed From Weeds', *Farmer's Weekly*, 30 August 1991, vol 115, part 9, pp 59–72.

9 Evans, J (1996) 'When to Remove Tubex Treeshelters – Notes from a Closely Observed Plantation', *Quarterly Journal of Forestry*, July 1996, vol 90, no 3, pp 207–208.

10 Nixon, C J (1994) 'Effectiveness of Tree Shelters in Upland Britain', *Quarterly Journal of Forestry*, January 1994, vol 88, no 1, pp 55–62.

11 *Developments in Treeshelter and Tree Guard Design: 1995*, Forestry Commission Research Division, Technical Development Branch, Technical Note 14/96.

12 Potter, M (1990) 'When Tree-shelters Break Down', *Heartwood*, NSWA newsletter July 1990, vol 2, p 5.

13 Pepper, H, Neil, D and Hemmings, J (1996) *Application of the Chemical Repellent Aaprotect to Prevent Winter Browsing*, Forestry Commission Research Information Note 289.

14 Forestry Authority (1996) *Controlling Grey Squirrel Damage to Woodlands*, Forestry Practice Advice Note 4, Edinburgh.

15 Pepper, H W (1990) *Grey Squirrel Damage Control with Warfarin*, Forestry Commission Research Note 180.

16 Agricultural Advisory and Development Service (1992) *Wild Rabbits*, Penwick, Northumberland.

17 Gill, R M A (1990) *Monitoring the Status of European and North American Cervids*, GEMS Information Series No 8, Global Environment Monitoring System, United Nations Environmental Programme, Nairobi, Kenya.

18 Harris, S, Morris, P, Wray, S and Yalden, D (1995) *A review of British mammals: population estimates and conservation status of British mammals other than cetaceans*, JNCC, Peterborough.

19 Forestry Authority (1995) *Managing Deer in the Countryside*, Forestry Practice Advice Note 2, Edinburgh.

20 Davies, R J and Gardiner, J B H (1987) *The Effect of Weed Competition on Tree Establishment*, Arboricultural Research Note 59/87, DoE, Arboricultural Advisory & Information Service.

21 (1996) *Bracken Whipping*, Forestry Commission, Technical Development Branch Technical Note 10/96, Dumfries.

22 Pengelly, D (1995) 'Dealing with a Sycamore Invasion', *Farming & Conservation*, October 1995, pp 32–34.

23 Willoughby, I and Dewar, J (1995) *The Use of Herbicides in the Forest*, Forestry Authority Field Book 8.

24 White, J and Patch, D (1990) *Ivy – Boon or Bane*, Arboricultural Advisory and Information Service, Research Note, Farnham.

25 Harris, E (1996) 'Management of Ivy', *Quarterly Journal of Forestry*, vol 90, no 2, April 1996, p 153.

26 (1996) *Standard Conditions of Contract and Specifications for Tree Works*, Arboricultural Association, Romsey, Hants.

27 Greig, J W and Strouts, R G (1983) *Honey Fungus*, Arboricultural Leaflet 2, Forestry Commission, HMSO, London.

28 Forestry Authority (1994) *Dutch Elm Disease in Britain*, Research Information Note 282.

29 Denyer, T (1994) *Dutch Elm Disease*, East Sussex State of Environment leaflet, East Sussex County Council.

30 Gibbs J, Brasier C, Webster J (1994) *Dutch Elm Disease in Britain*, Forestry Authority Research Information Note 252.

31 Redfern, D, Pratt, J and Whiteman, A (1994) *Stump Treatment against Fomes: A Comparison of Costs*, Forestry Authority Research Information Note 248.

CHAPTER 4

1 (1992) *Lowland Landscape Design Guidelines*, Forestry Authority, HMSO, London.

CHAPTER 5

1 Ferris-Kaan, R (1991) *Conservation Management of Woodlands by Non-Government Organisations*, Forestry Commission Research Note 199.

2 Elton, C S (1966) *The Pattern of Animal Communities*, Methuen, London.

3 Ferris-Kaan, R, Lonsdale, D and Winter, T (1993) *The Conservation Management of Deadwood in Forests*, Forestry Authority Research Information Note 241.

4 English Nature (1996) *Guide to the Care of Ancient Trees*, English Nature pamphlet, Peterborough.

5 Forestry Commission (1995) *Forest Operations and Badger Setts*, Forestry Commission Forest Practice Guide No 9, Edinburgh.

6 Forestry Commission (1976) *Badger Gates*, Forestry Commission Leaflet 68, HMSO, London.

7 Mayle, B A (1990) *Habitat Management for Woodland Bats*, Forestry Commission Research Information Note 165.

8 Rackham, O (1980) *Ancient Woodland: Its history, vegetation and uses in England*, Edward Arnold, London.

9 Biggs, J et al (1994) 'New Approaches to the Management of Ponds', *British Wildlife* vol 5, no 5, June, pp 273–287.

10 Biggs, J (1994) *New Approaches to Pond Management*, FWAG Bucks, Berks and Oxon Joint Newsletter.

11 Collinson, N H et al (1995) 'Temporary and Permanent Ponds: An Assessment of Drying Out on the Conservation Value of Aquatic Macroinvertebrate Communities', *Biological Conservation* 74, pp 125–133.

12 Biggs, J et al (1992) *Wetlands in Woodlands*, National Small Woods Association – Fourth Annual Conference Report, Preston Copes, Northamptonshire.

CHAPTER 7

1 (1979) *High Seats for Deer Management*, Forestry Commission leaflet 74, HMSO, London.

2 McPhail, D (1993) 'Woodlands and Wildlife Management', *Forestry & British Timber*, September 1993, pp 21–24.

APPENDIX 1

1 White, J E J (c1994) *The History of Introduced Trees in Britain*, Forestry Authority, Westonbirt Arboretum leaflet, Tetbury, Gloucestershire.

2 Rose, F (1976) 'Lichenological Indicators of Age and Environmental Continuity in Woodlands', in D H Brown et al (eds) *Lichenology: Progress and Problems*, The Systematics Association, Special Volume No 8, pp 279–307, Academic Press, London.

3 Mitchell, A F (1981) *The Native and Exotic Trees of Britain*, Arboricultural Research Note 29/81, DoE, Arboricultural Advisory & Information Service.

APPENDIX 3

1 (a) Cooper, C S (1995) 'Adverse Possession', *Timber Grower*, Autumn 1995, p 17.

(b) Cooper, C S (1996) 'Long Use of Rights of Way', *Timber Grower*, Spring 1996, p 27.

(c) Cooper, C S (1995) 'Check Your Access', *Timber Grower*, Spring 1995, p 23.

(d) Cooper, C S (1995) 'Contracts Made Over the Fence' *Timber Grower*, Summer 1995, p 22.

Index

Page numbers in *italics* refer to boxes, figures and tables. Those in **bold** refer to photographs.